SOLID STATE
DRIVES

As per the latest Anna University Syllabus
[For B.E., VI Semester EEE Students]

SOLID STATE DRIVES

As per the latest Anna University Syllabus
[For B.E., VI Semester EEE Students]

B. Balaji, Asst. Professor/ EEE,
IFET College of Engineering (Autonomous), Villupuram

Dr. P. Pugazhendiran, Professor/ EEE & COE
IFET College of Engineering (Autonomous), Villupuram

Dr. S. Saravanan, Professor/ EEE & Principal
CK College of Engineering & Technology, Cuddalore

R. Revathy, Assistant Professor/ EEE
IFET College of Engineering (Autonomous), Villupuram

ugam books
Powered by Noolus Publishing Pvt Ltd

ugam books

Puducherry | Chennai
Powered by Noolus Publishing Pvt Ltd

Solide State Drives by
B. Balaji, P. Pugazhendiran, S. Saravanan and R. Revathy

First published by **Ugam Books** 2020

ISBN 978-81-944825-3-6

Printed and Bound in India
ugam_011

For further information
w: www.ugambooks.com | www.noolus.com
m: ugambooks@gmail.com

Acknowledgement

First I thank my Parents **Mr. V. Baskaran Naidu & Mrs. B. Chithra** and my sister **Mrs. B. Bhuvaneswari** for their support, encouragement and cooperation to my career.

I imbrue my gratitude to my Guru's **Shri. VETHATHIRI MAHARISHI, Shri. CHANDRASEKARENDRA SARASWATHI SWAMIGAL** of **SHRI KANCHI KAMAKOTI PEETAM & SHIRIDI Shri. SAI BABA** for making me successful in all my endeavours.

I pay my sincere thanks to **Mr. K. V. Raja** – Chairman, **Mr. Shivram Alva** – Secretary, **Dr. G. Mahendran** – Principal, **Dr. S. Matilda** – Vice Principal & Dean Academics, **Dr. A. John Dhanaseely** – ASP/EEE & Head – IFET Research Centre, **Mr. M. Sujith** – ASP/EEE and all my colleagues of **IFET College of Engineering, Villupuram** for their valuable suggestions and guidance in making this book.

I pay my grateful thanks to my Co-Authors **Dr. P. Pugazhendiran & Mrs. R. Revathy** of IFET College of Engineering, Villupuram and **Dr. S. Saravanan** of CK College of Engineering & Technology, Cuddalore for making this book. I also thank **Mr. S. Gopinath & Ms. K. Yogapriya -** UG Scholar/EEE of IFET College of Engineering, Villupuram for helping us making this book.

I thank **Mr. V. Venogopalan Naidu** of **Annalakshmi Chitradevi Trust,** Puducherry and **Mr. A. Durairaj** of **Ugam Books**, Chennai for bringing out the book.

B. Balaji

Preface

We are pleased to bring out the book titled "SOLID STATE DRIVES" for VI Semester EEE students and the subject handlers. This book is based on the latest Anna University Syllabus. This book uses plain and lucid language to explain the fundamentals of this subject and provides very simple method for explaining various complicated concepts in stepwise manner.

Utmost care has been given while making this book to make students comfortable in understanding the basic concepts of the subject. Each units in this book are well supported with necessary illustration, examples and solved problems wherever necessary.

Any suggestions for the improvement of the book will be acknowledged and well appreciated. Paramount care and importance is borne especially to provide error free content, to all the beloved students across the world. Kindly excuse us, if you come across any errors anywhere in the book and let us know the same to rectify such things in future.

B. BALAJI

Dr. P. PUGAZHENDIRAN

Dr. S. SARAVANAN

R. REVATHY

Syllabus

Solid State Drives

(As Per Latest Anna University Syllabus)

	L	T	P	C
	3	0	0	3

UNIT I - DRIVE CHARACTERISTICS 9

Electric drive – Equations governing motor load dynamics – steady state stability – multi quadrant Dynamics: acceleration, deceleration, starting & stopping – typical load torque characteristics – Selection of motor.

UNIT II - CONVERTER / CHOPPER FED DC MOTOR DRIVE 9

Steady state analysis of the single and three phase converter fed separately excited DC motor drive– continuous conduction –Time ratio and current limit control – 4 quadrant operation of converter / chopper fed drive-Applications.

UNIT III - INDUCTION MOTOR DRIVES 9

Stator voltage control–V/f control– Rotor Resistance control-qualitative treatment of slip power recovery drives-closed loop control— vector control-Applications.

UNIT IV - SYNCHRONOUS MOTOR DRIVES 9

V/f control and self-control of synchronous motor: Margin angle control and power factor control - Three phase voltage/current source fed synchronous motor- Applications.

UNIT V - DESIGN OF CONTROLLERS FOR DRIVES 9

Transfer function for DC motor / load and converter – closed loop control with Current and speed feedback–armature voltage control and field weakening mode – Design of controllers; current controller and speed controller- converter selection and characteristics.

TOTAL: 45 PERIODS

Contents

UNIT 1
DRIVE CHARACTERISTICS

1.1. ELECTRICAL DRIVES - BASIC BLOCK DIAGRAM

✓ Drive is the combination of various systems combined together for the purpose of motion control.

✓ Drives that employs electric motors for the motion control is known as electrical drives.

✓ Electric Drives that employs solid state devices like power semiconductor switches for the control application is known as solid state drives.

✓ Applications of Electrical Drive: Paper mill, Cement Mill, cranes, lifts, etc.,

✓ Requirement of Electrical Drives,

   - Stable operation

   - Good transient response

BLOCK DIAGRAM OF ELECTRICAL DRIVES

a) Source

- Source would be 1Φ or 3Φ AC supply.

- Low power drives uses 1Φ supply and high power drives uses 3Φ supply system.

- Some drives are powered by battery with 24V, 48V and 110V DC.

b) Power Modulator

It is of 3 kinds,

i) Converters:

AC to DC Converters, AC Regulators, Choppers, Inverters and Cycloconverters can be used as converters.

ii) Variable Impedance:

   - Variable resistor, inductor is used to reduce the starting current.

- Liquid rheostat or slip regulator is used in case of high power applications.

iii) Switching circuits

- For changing the quadrant of operation.
- Operates the motor and drives in predetermined sequence.
- Prevents mall operations.
- Disconnects motor when there occurs abnormal operating conditions.

Functions of Power modulator

- Modulates the flow of power from source to motor.
- During transient operation, it makes the source and motor currents within the permissible values.
- Selects the mode of operation.
- Converts the electrical energy of source in the suitable form to the motor.

c) Electrical motors

Commonly used motors are,

DC Motors: Shunt motor, series motor, compound motor and permanent magnet motor.

AC Motors: Induction motor (squirrel cage, wound rotor & linear induction motor) & Synchronous motor (salient pole & smooth cylindrical).

Special Machines: Brushless DC motors, stepper motor & switched reluctance motor.

d) Sensing Unit

- Performs speed sensing and current sensing.
- ***Speed sensing*** – uses tachometer and it is known as closed loop speed control scheme.

- *Current sensing* – uses current sensor which uses Hall Effect and also uses non inductive resistance.

e) Control Unit

- Controls the power modulator.

- Consists of linear and digital integrated circuits and transistor to generate firing signals for the thyristors which is in power modulator.

- Also has microprocessor for the sophisticated control.

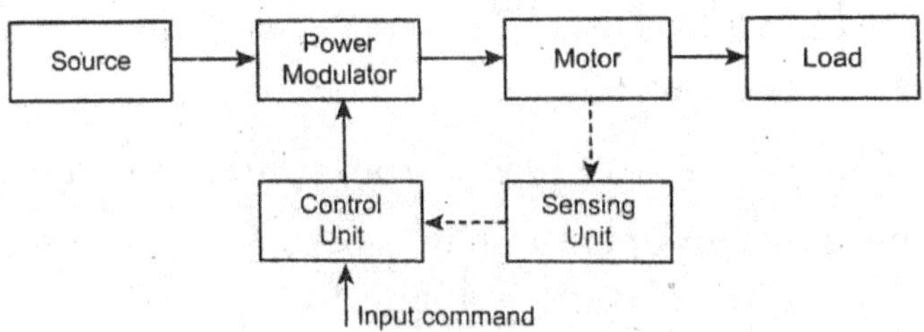

Fig.1.1. Electrical Drive – Block Diagram

1.2. KINDS OF ELECTRICAL DRIVES

a) Group Drive

- Single motor is used to drive an entire group of machine.

- Motor is connected to line shaft through belt and pulley.

- Used in earlier days, difficult to control and it is unsafe.

b) Individual Drive

- Has one motor for each working machine.

- Electrical motor is an internal part of the machine.

c) Multimotor Drive

- Has more than one motor for each working machine.

- Each motor is used to drive only one of the many working mechanism which is required in complicated production unit.

1.3. COMPARISON OF DC & AC DRIVES

DC Drives	AC Drives
Commutator makes the motor bulky, costly and heavy	Motors are inexpensive
Converter technology is well developed	Inverter technology is still being developed
Line commutation of converter is used	Forced commutation is used.
Fast response	Response depends on the type of control
Wide range of speed control	Wide range of speed control is only with solid state converters
Small power/ weight ratio	Large power/ weight ratio
Sparking at brushes	No sparking problems

1.4. EQUATION GOVERNING MOTOR LOAD DYNAMICS

- Motor drives a load through transmission system.

- When motor rotates, load may rotate or may undergo translation motion.

- Speed of the load may be different from motor speed, where it has many parts and it also have rotational motion or translation motion.

- It is convenient to have motor load system with equivalent rotational system.

Let,

J   – Polar moment of inertia of motor load (Kgm^2)

ω_m – Instantaneous angular velocity motor shaft (rad/sec)

T_m – Instantaneous value of developed motor torque (Nm)

T_l – Instantaneous value of load torque

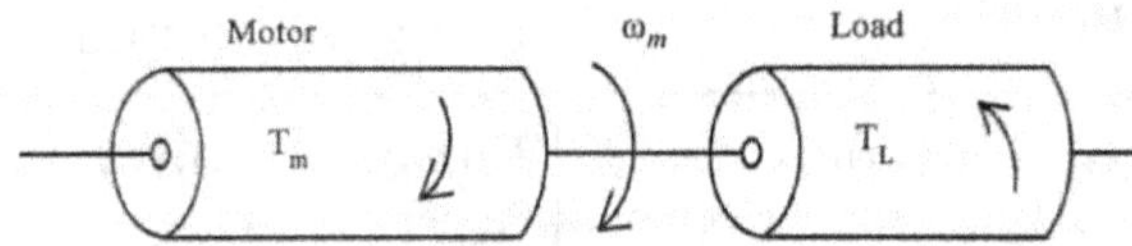

Fig.1.2. Motor Load System

Load torque includes friction and windage toque of motor.

For motor load system, fundamental torque equation,

$$T = \frac{d}{dt}(J\omega_m) \;----(1)$$

Take partial differentiation (1)

$$T = J.\frac{d\omega_m}{dt} + \omega_m.\frac{dJ}{dt} \;----(2)$$

Torque equation for motor load system will also be,

$$T = T_m - T_l \;----(3)$$

Where, T – Net Torque, T_m – Motor Torque &

    T_l – Load Torque

$$T_m - T_l = J.\frac{d\omega_m}{dt} + \omega_m.\frac{dJ}{dt} \;----(4)$$

Some drives will have variable inertia (J) (eg. Mine winder, industrial robot, etc.,)

Some drives will have constant inertial

In such case, $\dfrac{dJ}{dt} = 0$

So (4) becomes, $T_m - T_l = J.\dfrac{d\omega_m}{dt}$ ----(5)

$$T_m = T_l + J.\dfrac{d\omega_m}{dt} \text{ ----(6)}$$

WKT, T_m – Motor Torque, T_l – Load Torque &

$J.\dfrac{d\omega_m}{dt}$ – Dynamic Torque

So (6) shows, motor torque is counter balanced by load torque and dynamic torque.

Dynamic torque will be only during transient operation.

(5) => $T_m - T_l = J.\dfrac{d\omega_m}{dt}$

Case (i):   If $T_m > T_l$ – Drive experiences acceleration and speed increases.

Case (ii):   If $T_m < T_l$ – Drive experience deceleration motion and speed decreases.

Case (iii): If $T_m = T_l$ – Drive attains steady state and runs at constant speed.

Case (i) & (ii) is the transient process because acceleration and deceleration takes place. So the equations are called dynamic equations.

Components of Load Torque

Load Torque $(T_l) = T_f + T_w + T_L$

T_f *(Friction Torque):* Torque due to friction at motor shaft and various parts of the load.

T_w *(Windage Torque):* When motor runs, wind is generated and it opposes the rotating motion.

T_L *(Load Torque):* Torque required to do the useful mechanical work.

Motor Load System with translation motion

- Here the motor is connected to the load through the set of gears.

- Gears are used to amplify the torque on the load side but the speed would be low when compared to the motor speed.

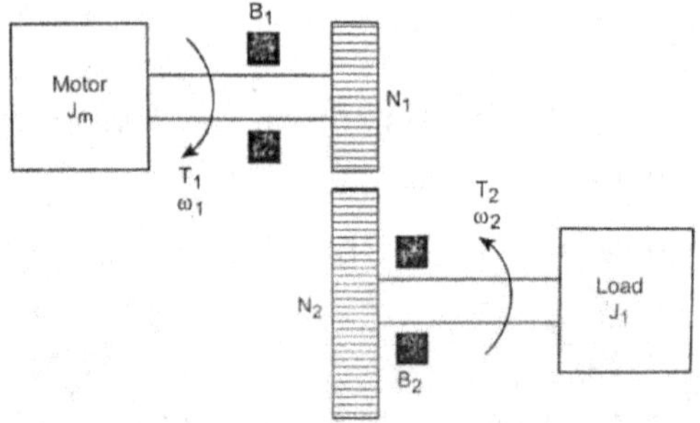

Fig.1.3. Motor load connection through gear

N_1, N_2 – Teeth number in the gear

B_1, B_2 – Bearings and their friction coefficient

J_1, J_m – Moment of inertia of the motor and load

Power handled by the gear in same on both sides,

So, $T_1\omega_1 = T_2\omega_2$ ----(1)

$$T_2 = T_1\left(\frac{\omega_1}{\omega_2}\right)\ ----(2)$$

Speed is inversely proportional to the number of teeth.

$$\frac{\omega_1}{\omega_2} = \frac{N_2}{N_1}\ ----(3)$$

Subs (3) in (2)

$$T_2 = T_1\left(\frac{N_2}{N_1}\right)\ ----(4)$$

The reflected inertia and bearing coefficients of load are,

$$J_1\ (\text{reflected}) = \left(\frac{N_1}{N_2}\right)^2 J_1 ----(5)$$

$$B_2 \text{ (reflected)} = \left(\frac{N_1}{N_2}\right)^2 B_2 \text{ ----(6)}$$

Resultant mechanical constants are,

$$J = J_m + \left(\frac{N_1}{N_2}\right)^2 J_1$$

$$B = B_1 + \left(\frac{N_1}{N_2}\right)^2 B_2$$

Torque equ below describes the motor load combination,

$$T_1 - T_2 \text{ (reflected)} = J.\frac{d\omega_{m1}}{dt} + \omega_{m1}.B$$

$$= T_1 - \left(\frac{N_1}{N_2}\right)T_2$$

1.5. MULTIQUADRANT DYNAMICS

- Motor operates in 2 modes – motoring and braking.
- Motoring – Converts electrical energy to mechanical energy (i.e) supports the motion.
- Braking – Converts mechanical energy to electrical energy (i.e) opposes the motion.
- Forward direction (or) upward motion – motor speed will be +ve – Acceleration.
- Reverse direction (or) downward motion – motor speed will be –ve – Deceleration.
- Hoist consists of a rope wound on a drum coupled to the motor shaft.
- One end of the rope is tied with a cage which is used to transfer the load and the other end of rope is tied with the counter weight.
- Weight of the counter weight is choosen to be higher than weight of empty cage but lower than a fully loaded cage.

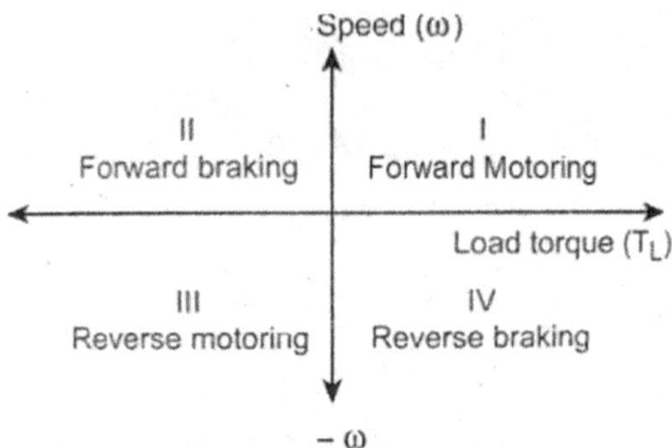

Fig.1.4. Four Quadrant operation

Quadrant I

- This operation comes when the cage is loaded.

- Here, the hoist requires upward motion and hence it is provided by the +ve motor speed.

- Upward motion with loaded cage will be obtained if $T_m=Tl_1$, the Tm should be positive.

- Here P is +ve and it is known as the forward motoring operation.

Quadrant II

- This operation comes when the cage is empty.

- The counter weight is heavier, than an empty cage, so that it pulls up the empty cage.

- In order to limit the speed within a safe value, motor must produce a braking torque equal to load torque Tl_2.

- Here P is –ve, ω_m is +ve and it is known as the forward braking operation.

Quadrant III

- This operation comes when the cage is empty.

- Here the empty cage is moved downwards, where an empty cage has lesser weight than the counter weight.

- Here the motor should produce a torque in clockwise direction.

- Speed is –ve, P is +ve and so it is known as reverse motoring operation.

Quadrant IV

- This operation comes when the cage is loaded.

- Weight of a loaded cage is higher than the counter weight. So that it would come down due to the gravity itself.

- In order to limit the speed of cage within a safe value, motor must produce +ve torque, $T_m = Tl_2$.

- Here P & ω is –ve, so it is known as the reverse braking operation.

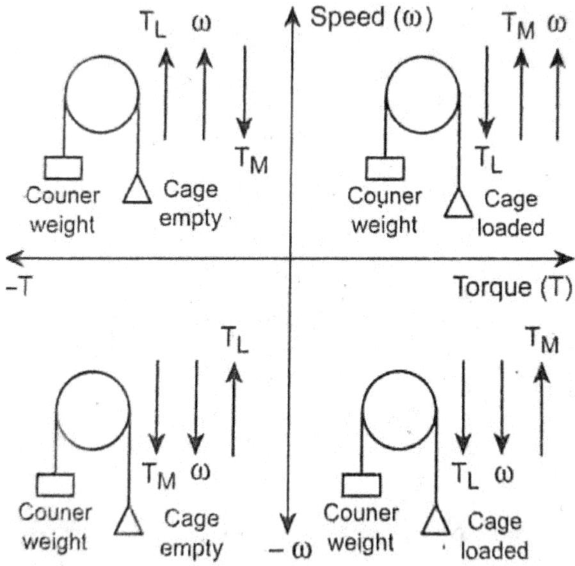

Fig.1.5. Four Quadrant Operation of a motor driving hoist

1.6. OPERATING MODES OF ELECTRICAL DRIVES (STEADY STATE OPERATION, ACCELERATION & DECELERATION)

STEADY STATE OPEARATION

- Steady state operation takes place when the motor torque T_m is equal to T_l.

- When the motor parameters are adjusted to provide speed torque curve 1, drive runs at speed ω_{m1}.

- Speed is changed to ω_{m2}, when the motor parameters are adjusted to provide speed torque curve 2.

- If the loaded hoist is to be lowered or unloaded hoist is to be lifted, the load torque assists the motion.

- To have a steady state operation, mechanical brake is used to oppose the motion.

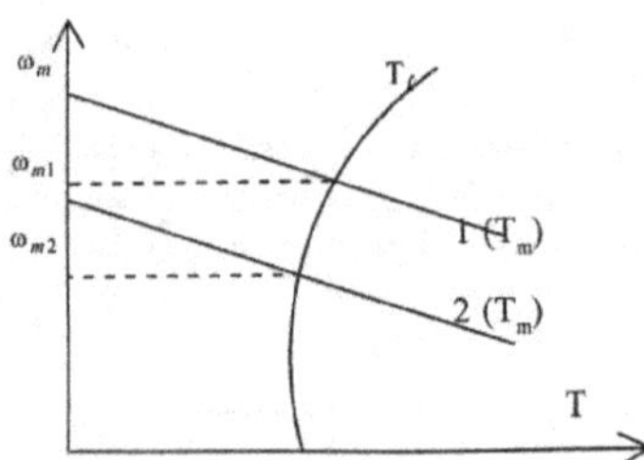

Fig.1.6. Principle of speed control

ACCELERATION

- Acceleration is done, whenever an increase in speed is required.

- Acceleration can be made by increasing the motor torque than the load torque.

- Increase in motor torque increases the motor current, thus it is to be within the safer value.

- Here the transition occurs from point A at speed ωm_1 to point B at speed ωm_2 but the torque remains constant.

- Path of acceleration is A-D_1-E_1-B.

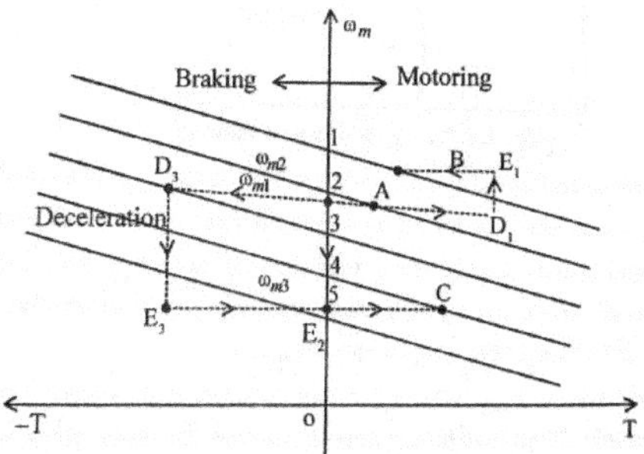

Fig.1.7. Acceleration & Deceleration

Soft Start: In some applications the motor should accelerate smoothly without any jerk. This is achieved by increasing the starting torque steplessly from its zero value. This type of start is known as soft start.

DECELERATION

- Deceleration is done, whenever decrease in speed is required.

- Deceleration can be made by decreasing the motor torque than the load torque.

- Deceleration is done by electric braking at constant braking torque.

- Here the transition occurs from point A at speed ωm1 to point C at speed ωm3.

- Path of deceleration is A-D₃-E₃-E₂-C.

STARTING

- If the motor is started with full voltage across the terminals, current will be in order of 20 times the rated current.

- Such high current will damage the motor. So the current has to be limited within the safer value during starting.

- This is achieved by decreasing the source voltage by connecting resistance in series with the motor terminal.

- Controller is also provided for the control of speed by limiting the starting current.

STOPPING (BRAKING)

- During braking, the machine works as a generator.

- Machine will develop a positive torque during the time of starting (machine acts as a motor) and machine develops a negative torque during the time of braking (machine acts a generator).

- There are 3 types of braking,

   a) *Plugging:* Here, in order to brake the motoring operation, the supply voltage is reversed. So that, negative torque is generated, this opposes the positive torque.

   b) *Dynamic Braking:* Here, the motor armature is disconnected from the source and connected across the resistance. So that, the motor works as a generator , producing the braking torque.

   c) *Regenerative Braking:* Here, the energy generated is supplied to the source itself. Source does not have ability to store the energy, so the energy is diverted to load connected to the source. In this braking, the induced emf should be greater than the supply voltage.

1.7. STEADY STATE STABILITY

- Dynamic conditions occur in electric drives during transient operations.

- Dynamic condition also arises when there is variable speed.

- The system is said to be stable, even after the disturbance, the drive attains the equilibrium position.

- At constant speed, if the developed motor torque is equal to the sum of load torque and friction – Equilibrium state (no disturbance).

- The influencing factor of stability,
    a) Inertia of rotating motor (J)
    b) Inductance of motor (L)

- If the changes from one state of equilibrium to another takes place too slowly to have the effects of the above factors, the stability condition refers to steady state stability.

- Motor load system obtains equilibrium speed, when $T_m = T_l$ and this is the speed at which the drive will operate in steady state.

- Let us have the steady state of equilibrium point as A and dynamic condition as the point B.

- In point A, even though there is a disturbance, the drive will restore its equilibrium position.

- In point B, if there is a disturbance, the drive would decelerate ($T_m < T_l$) or accelerate ($T_m > T_l$).

- Let a small perturbation in speed $\Delta\omega_m$ and ΔT_l.

$$T_m = T_l + \frac{J d\omega_m}{dt} \quad ----(1)$$

$$(T_m + \Delta T_m) = (T_l + \Delta T_l) + J.\frac{d}{dt}(\omega_m + \Delta\omega_m) \quad ----(2)$$

Subtracting (1) & (2)

$$T_m + \Delta T_m - T_m = T_l + \Delta T_l - T_l + J.\frac{d\omega_m}{dt} + J.\frac{d\Delta\omega_m}{dt} - J.\frac{d\omega_m}{dt}$$

$$\Delta T_m = \Delta T_l + J.\frac{d\Delta\omega_m}{dt}$$

$$J.\frac{d\Delta\omega_m}{dt} + \Delta T_l - \Delta T_m = 0 \quad --(3)$$

- For small perturbation, the speed torque curves of motor and load is assumed to straight line.

So,

$$\Delta T_m = \left(\frac{dT_m}{d\omega_m}\right)\Delta\omega_m \quad ----(4)$$

$$\Delta T_l = \left(\frac{dT_l}{d\omega_m}\right)\Delta\omega_m \quad ----(5)$$

Sub (4) & (5) in (3)

$$J.\frac{d\Delta\omega_m}{dt} + \left(\frac{dT_l}{d\omega_m} - \frac{dT_m}{d\omega_m}\right)\Delta\omega_m = 0 \quad ----(6)$$

Characteristic equation for (6) will be,

$$(J.D+C)\Delta\omega_m = 0$$

$$\left(D + \frac{C}{J}\right)\Delta\omega_m = 0$$

Taking complementary function for above equ.

$$m + \frac{C}{J} = 0 \Rightarrow m = -\frac{C}{J}$$

Real roots, $\Delta\omega_m = Ae^{mt}$

so, $\Delta\omega_m = Ae^{-\frac{C}{J}t}$

$$\Delta\omega_m = Ae^{-\frac{1}{J}\left(\frac{dT_l}{d\omega_m} - \frac{dT_m}{d\omega_m}\right)t} \quad ----(7)$$

At initial condition, t=0, (7) becomes,

$$\Delta\omega_m = Ae^0 \Rightarrow \Delta\omega_m = A$$

Assume $A = \Delta\omega_{m0}$

$$\Delta\omega_m = (\Delta\omega_{m0})\, e^{-\frac{1}{J}\left(\frac{dT_l}{d\omega_m} - \frac{dT_m}{d\omega_m}\right)t}$$

Operation will be stable, when $A\omega_m$ approaches zero and t approaches infinity.

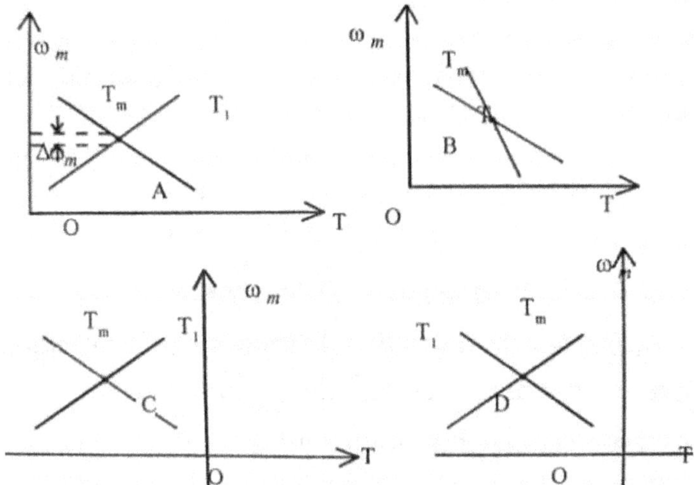

Fig.1.8. Points A&C are stable and B&D are unstable

1.8. TYPICAL LOAD TORQUE CHARACTERISTICS

- The main factor in selecting the type of motor for the driving the load mainly depends upon the speed torque characteristics.

- Different loads will have different speed torque characteristics.

- Loads are classified into,

    a) Constant torque type (T=K)

    b) Generator type (T$\alpha\omega$)

    c) Fan type (T$\alpha\omega^2$)

    d) Constant power type (T$\alpha\dfrac{1}{\omega}$)

a) Constant torque type

- Here the torque is constant irrespective of the speed.

- Most of the machines which has mechanical nature of working like shaping, cutting, grinding etc., uses constant torque type.

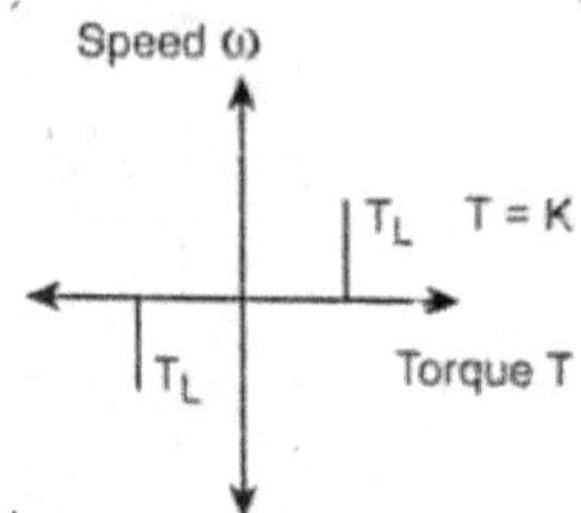

Fig.1.9. Constant torque type

- Cranes during hoisting and conveyors during handling the weight also uses constant torque type.

- Characteristic Equation: T=K

b) Generator type

- Here the torque is directionally proportional to the speed.

- Separately excited DC generator connected with constant resistance load and calendaring machines comes under this type of load.

- Characteristic equation: T=Kω.

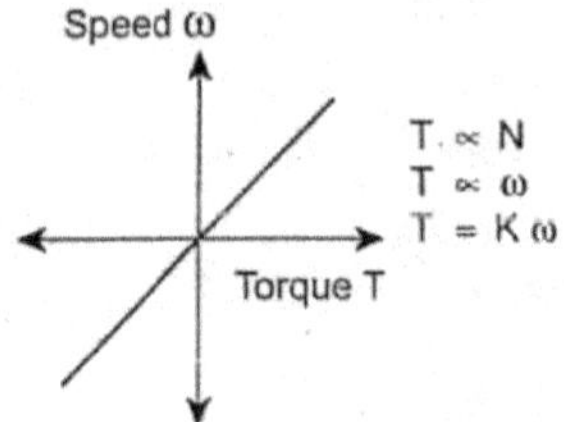

Fig.1.10. Generator Type   *Fig.1.11. Fan Type*

c) Fan type

- Here the torque is proportional to the square of the speed.

- Fan, rotary pumps, compressors etc., comes under this type.

- Characteristic equation: $T=K\omega^2$

d) Constant power type

- Here the torque is inversely proportional to the speed and the load power is constant.

- Certain type of lathes, boring machines, milling machines, steel mill coilers etc., comes under constant power type.

- Characteristic curve will be in hyperbolic form.

- Characteristic equation: $T = K . \dfrac{1}{\omega}$

- This inverse proportionality is also due to the breakaway torque.

- Some machines need extra effort (torque) at the time of starting to overcome static friction. Such torque is breakaway torque and it is known as stiction.

- Due to this stiction, there is change in speed, near the zero speed.

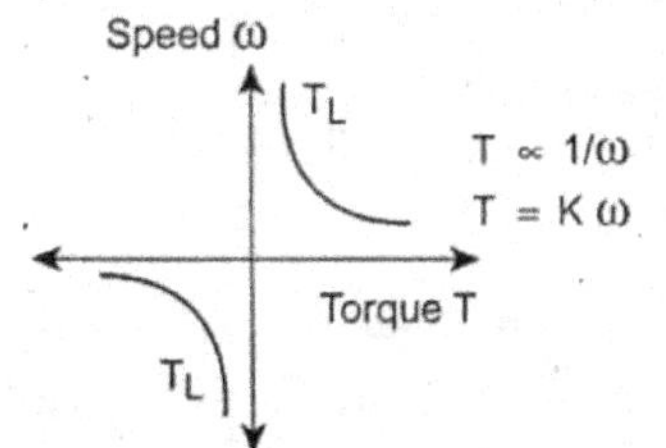

Fig.1.12. Constant Power Type

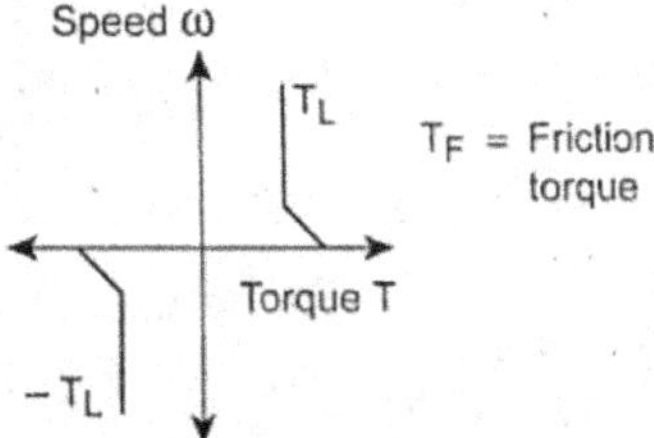

Fig.1.13. Constant Type load-Stiction

1.9. SELECTION OF MOTOR

- Power rating for the specific application must be chosen carefully to achieve reliability.

- Choosing insufficient rating, reduces the reliability. It also leads to extra initial cost and high loss of energy.

- The selection of motors for the specific applications are influenced by the following factors,

- ▪ Initial and running costs
- ▪ Type of power supply
- ▪ Speed – torque characteristics
- ▪ Torque capability
- ▪ Power density
- ▪ Thermal capacity
- ▪ Robustness
- ▪ Nature of environment, etc.,
- Electrical machines which are used for control applications are,
  - ▪ *DC Machines* – Shunt, series, compound and separately excited DC motor.
    - DC machines are expensive and requires periodic maintenance
    - It is fast, accurate and continuous speed control over a wide range,
    - Used in rolling mills, paper mills, electrical traction, cranes, hoists, etc.,
  - ▪ *AC Machine* – Induction motor (squirrel cage & slip ring type), synchronous motor and permanent magnet synchronous motor.
    - Squirrel cage type is least expensive, robust and maintenance free but it has poor power factor, low starting torque etc.,
    - Used in fans, pumps, conveyors etc.,
    - Slip ring type is more expensive, has higher starting toque. Can be started and stopped frequently as required.
    - Used in hoists, conveyors, elevators, etc.,

- Synchronous motor has higher initial cost, low running cost.

- Used to drive loads where comfort speed is needed.

- **Special Machines** – Stepper motor and switched reluctance motor.

    - Stepper motor converts digital signals into mechanical output.

    - Used in line printers, floppy disc drives and also used for mixing, stirring, cutting, etc.,

SOLVED PROBLEMS

1. **A motor drives two loads. One has rotational motion. It is coupled to the motor through a reduction gear with a=0.1 and efficiency of 90%. The load has a moment of inertia of 10Kgm² and a torque of 10 NM. Other load has a translation motion and consists of 1000Kg weight to be lifted up at a uniform speed of 85%. Motor has inertia of 0.2 Kg-m² and runs at a constant speed of 1420 rpm. Determine equivalent inertia referred to the motor shaft and power developed by the motor.**

Solution,

$$J = J_0 + a_1^2 J_1 + M_1\left(\frac{V_1}{\omega_m}\right)^2$$

$J_0 = 0.2$ Kg-m², $a_1=0.1$, $J_1=10$Kg-m²

$n = 1.5$ m/s

$$\omega_m = \frac{1420*\pi}{30} = 148.7 \text{ rad/s}$$

$$J = 0.2 + (0.1)^2*10 + 1000\left(\frac{1.5}{148.7}\right)^2 = 0.4 \text{ Kg-m}^2$$

$$T_L = \frac{a_1 T_{L1}}{n_1} + \frac{F_1}{\eta_1^1}\left(\frac{V_1}{\omega_m}\right)$$

$$\eta_1 = 0.9, \ \eta_1^l = 0.1$$

$$T_{L1} = 10Nm, \ \eta_1^l = 0.85$$

$$F_1 = 1000*9.81N, \ n_1 = 1.5 \ m/s$$

$$\omega_m = 148.7 \ rad/s$$

$$T_L = \frac{0.1*10}{0.9} * \frac{1000*9.81}{0.85} \left(\frac{1.5}{148.7}\right)$$

$$T_L = 117.53 \ Nm$$

Power developed $= T_L\omega_m = 117.53*148.7$

$$P = 17.48KW$$

UNIT 2
CONVERTER/CHOPPER FED DC MOTOR DRIVE

2.1. STEADY STATE ANALYSIS OF SINGLE PHASE FULL CONVERTER FED SEPARATELY EXCITED DC MOTOR DRIVE

✓ This drive is used for the control of low and medium horse power applications.

✓ Here the load is the equivalent circuit consisting of armature circuit resistance (R_a), armature circuit inductance (L_a) and back emf (E_b).

✓ This equivalent circuit is connected with the bridge circuit consisting of thyistors T_1, T_2, T_3 and T_4.

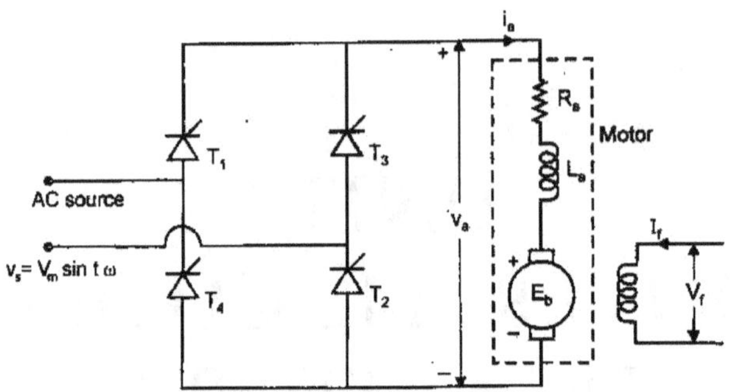

Fig.2.1. Single Phase full converter fed DC Drive

✓ AC input voltage V_s is applied to the bridge circuit.

$V_s = V_m \sin \omega t$

✓ This converter circuit operates in two modes. Continuous conduction mode and discontinuous conduction mode.

a) Continuous conduction mode

- When the armature current flows continuously, the conduction is said to be continuous mode of conduction.

- During positive half cycle, T_1&T_2 is ON and T_3&T_4 is OFF and during negative half cycle T_3&T_4 is ON and T_1&T_2 is OFF.

- T_1&T_2 is turned ON during positive half cycle from angle $\omega t=\alpha$ to angle $\omega t=\pi$.

- From angle $\omega t=\pi$ to $\omega t=\pi+\alpha$, no thyristors conducts.

- T_3&T_4 is turned ON during negative half cycle from angle $\omega t=\pi+\alpha$ to angle $\omega t=2\pi$.

- When supply voltage appears across the thyristors T_1&T_2, it is forward biased and gets turned ON. Orelse, T_1&T_2 are reverse biased, they gets turned OFF. This process is known as natural or line commutation.

- During α to π, energy flows from the input supply to the motor and during π to $\pi+\alpha$, some of the motor energy is fedback to the input supply.

- When the thyristors gets turned ON, the motor load is connected with the source Vs.

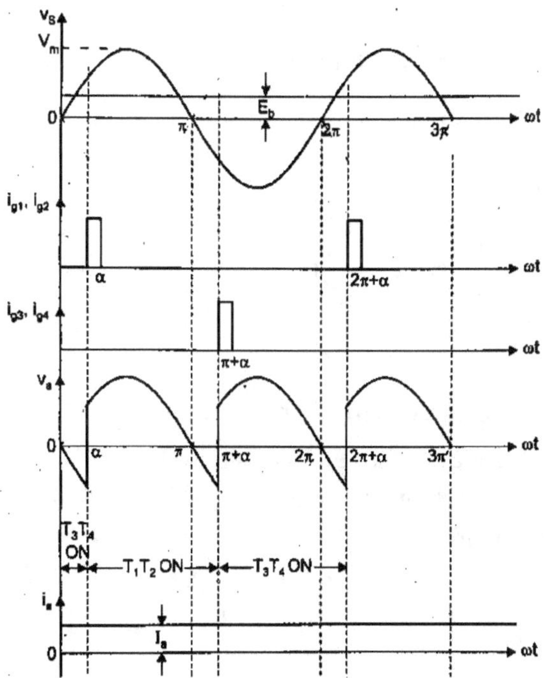

Fig.2.2. Output Waveform (Continuous Conduction)

$$V_a = \frac{1}{\pi} \int_{\alpha}^{\pi+\alpha} V_m \sin \omega t \, d\omega t$$

$$= \frac{V_m}{\pi} \int_{\alpha}^{\pi+\alpha} \sin \omega t \, d\omega t$$

$$= \frac{V_m}{\pi} [-\cos \omega t]$$

$$= \frac{V_m}{\pi}[-\cos(\pi+\alpha)+\cos \alpha]$$

$$V_a = \frac{2V_m}{\pi} \cos \alpha \quad \text{-----(1)}$$

WKT,

$$T = KI_a \quad \text{----(2)}$$

$$E = K\omega_m \quad \text{----(3)}$$

$$V_a = E + I_a R_a \quad \text{----(4)}$$

Equating (1) & (4)

$$\frac{2V_m}{\pi} \cos \alpha = E + I_a R_a \quad \text{----(5)}$$

Subs (2) & (3) in (5)

$$\frac{2V_m}{\pi} \cos \alpha = K\omega_m + \frac{T}{K}R_a$$

$$K\omega_m = \frac{2V_m}{\pi} \cos \alpha - \frac{T}{K}R_a$$

$$\omega_m = \frac{2V_m}{K\pi} \cos \alpha - \frac{T}{K^2}R_a$$

b) Discontinuous conduction mode

- When the armature current does not flow continuously, the conduction is said to be discontinuous mode of conduction.

- During positive half cycle, T_1&T_2 is ON and T_3&T_4 is OFF and during negative half cycle T_3&T_4 is ON and T_1&T_2 is OFF.

- T_1&T_2 is turned ON during positive half cycle from angle $\omega t = \alpha$ to angle $\omega t = \pi$.

- From the angle ωt=π to some angle ωt=β called extinction angle, the load current reduced to zero.

- From angle ωt=β to ωt=π+α, no thyristors conducts. so the load current is zero but there is voltage which is equal to E..

- T₃&T₄ is turned ON during negative half cycle from angle ωt=π+α to angle ωt=2π.

- When the thyristors gets turned ON, the motor load is connected with the source Vs.

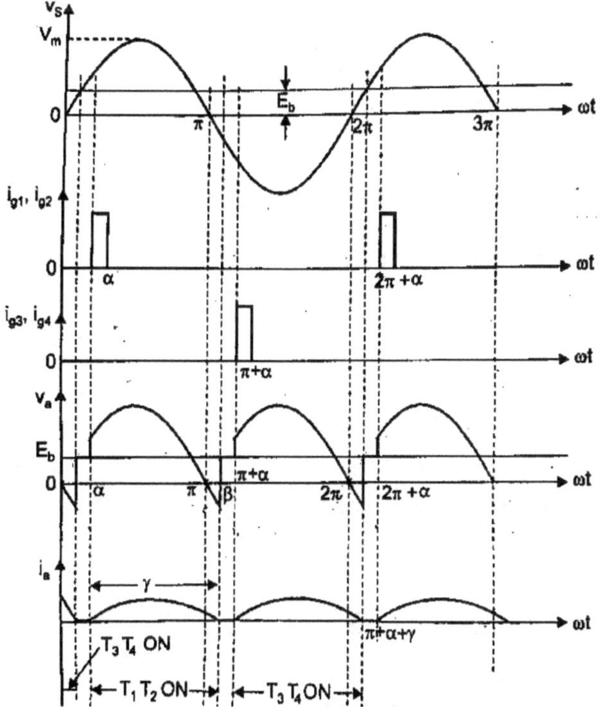

Fig.2.3. Output Waveform (Discontinuous Conduction)

$$V_a = \frac{1}{\pi} \left(\int_{\alpha}^{\beta} V_m \, \mathrm{Sin}\, \omega t \, dwt + \int_{\beta}^{\pi+\alpha} E \, dwt \right)$$

$$= \frac{1}{\pi} \left(V_m[-\mathrm{Cos}\, \omega t] + E[\omega t] \right)$$

$$= \frac{1}{\pi}\left(V_m[-\text{Cos }\beta + \text{Cos}\alpha] + E[\pi+\alpha-\beta]\right)$$

$$= \frac{1}{\pi}\left(V_m[\text{Cos}\alpha - \text{Cos }\beta] + E[\pi+\alpha-\beta]\right)$$

$$= \frac{V_m}{\pi}[\text{Cos}\alpha - \text{Cos }\beta] + \frac{E}{\pi}[\pi+\alpha-\beta] \quad ----(1)$$

WKT,

$$V_a = E + I_a R_a \quad ----(2)$$

$$T = K I_a \quad ----(3)$$

$$E = K\omega_m \quad ----(4)$$

From (2), $E = V_a - I_a R_a \quad ----(5)$

Subs (3) & (4) in (5)

$$K\omega_m = V_a - \frac{T}{K}R_a$$

$$\omega_m = \frac{V_a}{K} - \frac{T}{K^2}R_a \quad ----(6)$$

Subs (1) in (6)

$$\omega_m = \frac{V_m(\text{Cos }\alpha - \text{Cos }\beta)}{K.\pi} + \frac{E(\pi+\alpha-\beta)}{K.\pi} - \frac{T}{K^2}R_a \quad ----(7)$$

WKT,

$$\beta = \alpha + \pi \Rightarrow \pi = \beta - \alpha \quad ----(8)$$

Subs (8) in (7)

$$\omega_m = \frac{V_m(\text{Cos }\alpha - \text{Cos }\beta)}{K(\beta-\alpha)} + \frac{E(\pi+\alpha-\alpha-\pi)}{K.\pi} - \frac{T.R_a}{K^2}$$

$$\omega_m = \frac{V_m(\text{Cos }\alpha - \text{Cos }\beta)}{K(\beta-\alpha)} - \frac{T.R_a}{K^2}$$

$$\omega_{mc} = \frac{R_a}{KZ} V_m \text{Sin }(\alpha-\Phi) \frac{(1+e^{-\pi\,\text{Cot }\emptyset})}{(e^{-\pi\,\text{Cot }\emptyset}-1)}$$

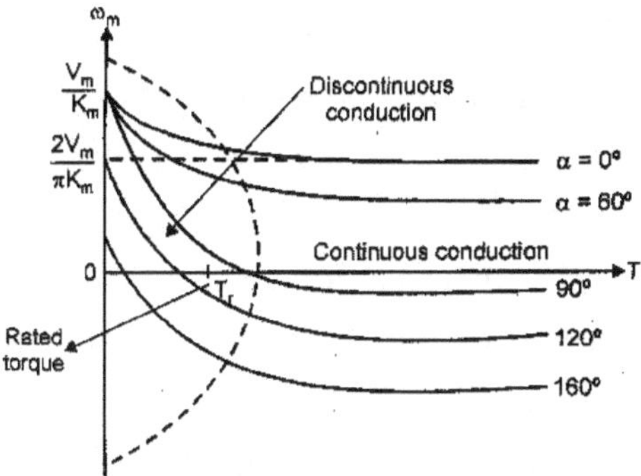

Fig.2.4. Torque Speed Characteristics

2.2. STEADY STATE ANALYSIS OF SINGLE PHASE HALF CONVERTER FED SEPARATELY EXCITED DC MOTOR DRIVE

✓ Converter circuit consists of 2 thyristors T_1 & T_2, 2 diodes D_1 & D_2 and the load is an equivalent circuit consisting of armature circuit resistance (R_a), armature circuit inductance (L_a) and back emf (E_b).

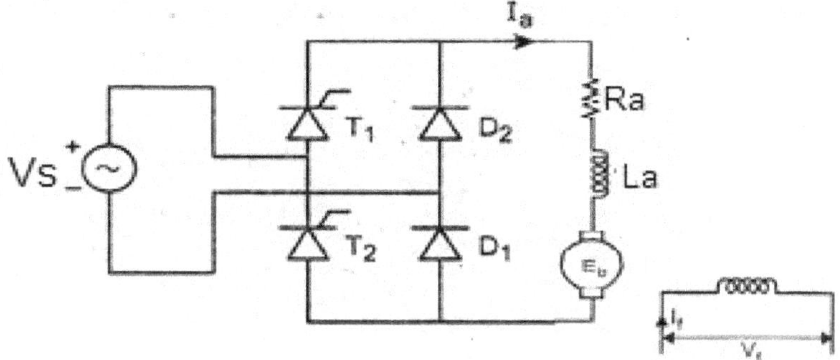

Fig.2.5. Single Phase half converter fed DC Drive

✓ AC input voltage Vs is applied to the bridge circuit.

$$V_s = V_m \, Sin \, \omega t$$

✓ This converter circuit operates in two modes. Continuous conduction mode and discontinuous conduction mode.

a) Continuous conduction mode

- When the armature current flows continuously, the conduction is said to be continuous mode of conduction.

- During positive half cycle, T_1&D_1 is ON and T_2&D_2 is OFF and during negative half cycle T_2&D_2 is ON and T_1&D_1 is OFF.

- T_1&D_1 is turned ON during positive half cycle from angle $\omega t=\alpha$ to angle $\omega t=\pi$.

- From angle $\omega t=\pi$ to $\omega t=\pi+\alpha$, no thyristors and diodes conducts and the armature current freewheels through the diodes D_1&D_2.

- T_2&D_2 is turned ON during negative half cycle from angle $\omega t=\pi+\alpha$ to angle $\omega t=2\pi$.

- When the thyristors and diodes gets turned ON, the motor load is connected with the source Vs.

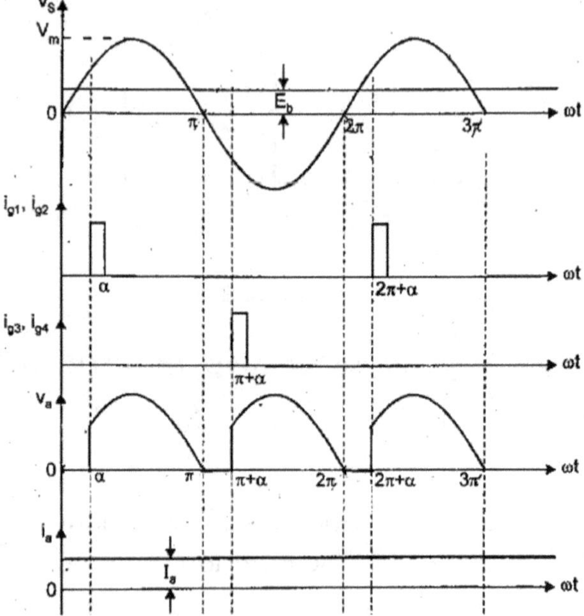

Fig.2.6. Output Waveform (Continuous Conduction)

$$V_a = \frac{1}{2\pi} \int_\alpha^\pi V_m \, Sin \, \omega t \, d\omega t$$

$$= \frac{V_m}{2\pi} [- Cos \, \omega t]$$

$$= \frac{V_m}{2\pi} [- Cos \, \pi + Cos \, \alpha]$$

$$= \frac{V_m}{2\pi} [1 + Cos \, \alpha] \ ---- (1)$$

WKT,

$$T = K I_a \ ---- (2)$$

$$E = K \omega_m \ ---- (3)$$

$$V_a = E + I_a R_a \ ---- (4)$$

Equating (1) & (4)

$$\frac{V_m}{2\pi} [1 + Cos \, \alpha] = E + I_a R_a \ ---- (5)$$

Subs (2) & (3) in (5)

$$\frac{V_m}{2\pi} [1 + Cos \, \alpha] = K \omega_m + \frac{T}{K} R_a$$

$$K \omega_m = \frac{V_m}{2\pi} [1 + Cos \, \alpha] - \frac{T}{K} R_a$$

$$\omega_m = \frac{V_m}{2K\pi} [1 + Cos \, \alpha] - \frac{T}{K^2} R_a$$

b) Discontinuous conduction mode

- When the armature current does not flow continuously, the conduction is said to be discontinuous mode of conduction.

- During positive half cycle, T_1&D_1 is ON and T_2&D_2 is OFF and during negative half cycle T_2&D_2 is ON and T_1&D_1 is OFF.

- T_1&D_1 is turned ON during positive half cycle from angle $\omega t = \alpha$ to angle $\omega t = \pi$.

- From the angle $\omega t=\pi$ to some angle $\omega t=\beta$ called extinction angle, the load current reduced to zero.

- From angle $\omega t=\beta$ to $\omega t=\pi+\alpha$, no thyristors conducts so the load current is zero but there is voltage which is equal to E.

- T_2&D_2 is turned ON during negative half cycle from angle $\omega t=\pi+\alpha$ to angle $\omega t=2\pi$.

- When the thyristors gets turned ON, the motor load is connected with the source Vs.

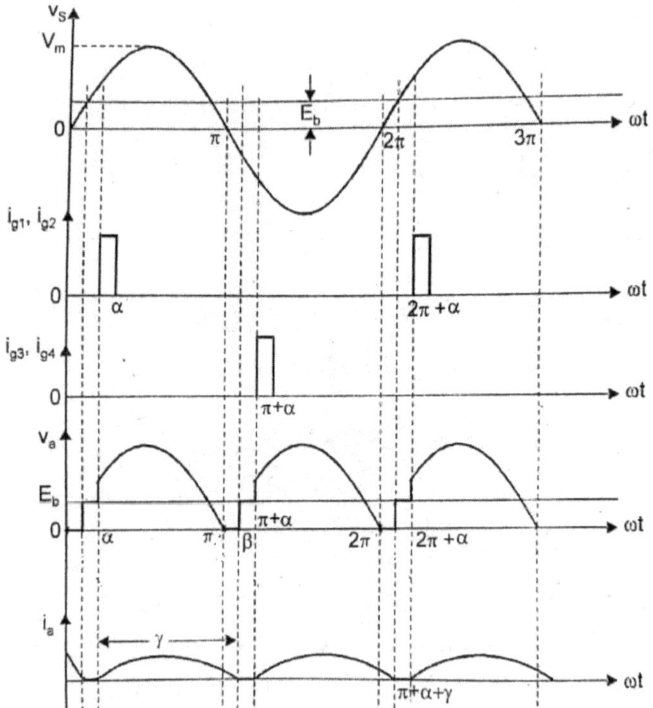

Fig.2.7. Output Waveform (Discontinuous Conduction)

$$V_a= \frac{1}{2\pi}[\int_\alpha^\pi V_m \sin \omega t\, d\omega t+\int_\pi^\beta 0\, d\omega t +\int_\beta^{\pi+\alpha} E\, d\omega t$$

$$V_a = \frac{1}{2\pi}[\int_\alpha^\pi V_m \sin \omega t\, d\omega t +\int_\beta^{\pi+\alpha} E\, d\omega t]$$

$$V_a = \frac{V_m}{2\pi}[-\cos \omega t] + \frac{E}{2\pi}[\omega t]$$

Subs. the upper & lower limits

$$= \frac{V_m}{2\pi}[-\text{Cos } \pi + \text{Cos } \alpha] + \frac{E}{2\pi}[\pi+\alpha-\beta]$$

$$= \frac{V_m}{2\pi}[1+ \text{Cos } \alpha] + \frac{E}{2\pi}[\pi+\alpha-\beta] \quad ----(1)$$

WKT,

$$T=KI_a \quad ----(2)$$

$$E=K\omega_m \quad ----(3)$$

$$V_a = E+I_aR_a \quad ----(4)$$

Subs (1), (2) & (3) in (4)

$$\frac{V_m}{2\pi}[1+ \text{Cos } \alpha] + \frac{E}{2\pi}[\pi+\alpha-\beta] = K\omega_m + R_a.\frac{T}{K}$$

$$K\omega_m = \frac{V_m}{2\pi}[1+ \text{Cos } \alpha] + \frac{E}{2\pi}[\pi+\alpha-\beta] - R_a.\frac{T}{K}$$

$$\omega_m = \frac{V_m}{2\pi.K}[1+ \text{Cos } \alpha] + \frac{E}{2\pi.K}[\pi+\alpha-\beta] - R_a.\frac{T}{K^2} \quad ----(5)$$

WKT, $\beta=\pi+\alpha => \pi=\beta-\alpha$ ----(6)

Subs (6) in (5)

$$\omega_m = \frac{V_m}{2K(\beta-\alpha)}[1+ \text{Cos } \alpha] + \frac{E}{\pi.K}[\pi+\alpha-\pi-\alpha] - R_a.\frac{T}{K^2}$$

$$\omega_m = \frac{V_m}{2K(\beta-\alpha)}[1+ \text{Cos } \alpha] - R_a.\frac{T}{K^2}$$

$$\omega_{mc} = \frac{V_m}{2K}.\frac{R}{Z}.\frac{[\text{Sin } \emptyset \, e^{-\alpha \, \text{Cot} \emptyset}-\text{Sin } (\alpha-\emptyset)e^{-\pi \, \text{Cot} \emptyset}]}{1- e^{-\pi \, \text{Cot} \emptyset}}$$

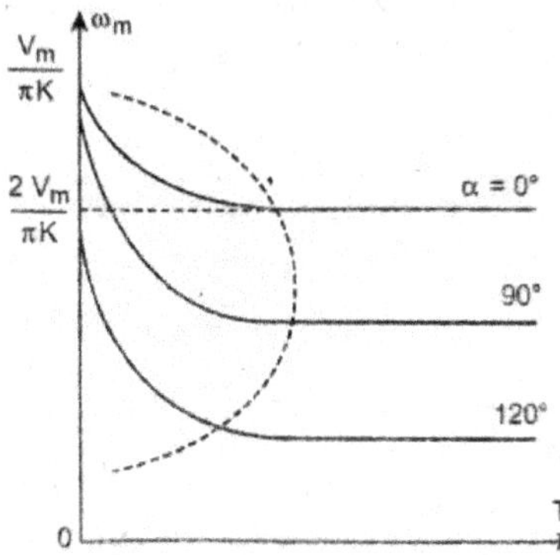

Fig.2.8. Torque Speed Characteristics

2.3. STEADY STATE ANALYSIS OF THREE PHASE FULL CONVERTER FED SEPARATELY EXCITED DC MOTOR DRIVE

- It is used in industrial applications upto 1500 KW drives.

- Here, average voltage output will be either positive or negative but the average current output will always be positive.

- Three phase AC supply is connected to the input terminals of converter circuit and the armature terminals of the motor are connected with the output terminals of the converter circuit.

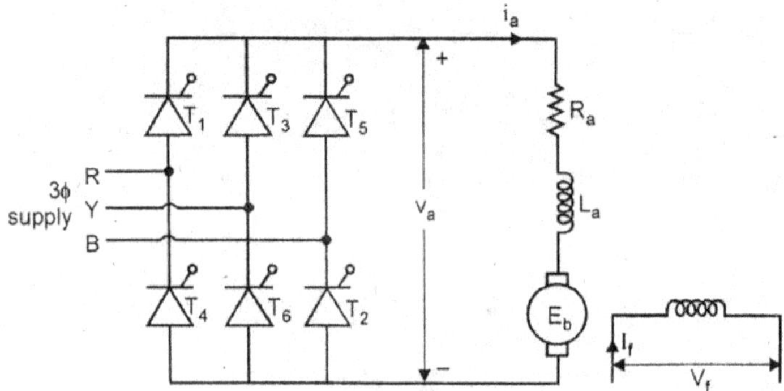

Fig.2.9. Three Phase Full converter fed DC Drive

- Converter circuit works as three phase AC to DC converter for firing angle $0^0 < \alpha < 90^0$ and the circuit works as three phase line commutated inverter for the firing angle $90^0 < \alpha < 180^0$.

- Converter circuit consists of 6 thyristors T_1, T_2, T_3, T_4, T_5 & T_6. In this, T_1, T_3 & T_5 are positive group of thyristors and T_2, T_4 & T_6 are negative group of thyristors.

- Each thyristor conducts for 120^0 and the thyristor pair conducts for 80^0.

At, $\alpha < 90^0$

- ✓ For $\alpha = 0^0$, all the thyristors behaves like a diodes. Here, T_1 is fired at $\omega t = 60^0$, T_2 at $\omega t = 90^0$, T_3 at 150^0 and so on.

- ✓ For $\alpha = 60^0$, T_1 is fired at $\omega t = 90^0$, T_2 at $\omega t = 120^0$ and so on.

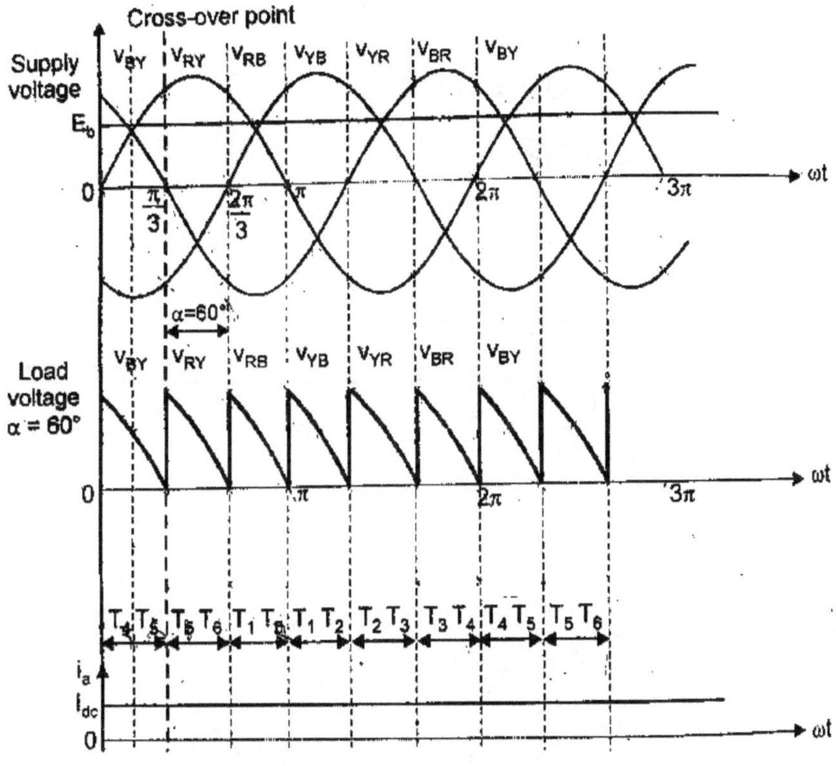

Fig.2.10. Output Waveform ($\alpha < 900$)

At, $\alpha > 90^0$

- ✓ For $\alpha = 120^0$, here the motor terminal voltage becomes negative so that it is known as inversion mode of operation.

- ✓ The power flow occurs from load to source.

$$V_d = \frac{3}{\pi} \int_{\alpha + \frac{\pi}{3}}^{\alpha + \frac{2\pi}{3}} V_m \, \text{Sin } \omega t \, d\omega t$$

$$= \frac{3V_m}{\pi} \int_{\alpha+\frac{\pi}{3}}^{\alpha+\frac{2\pi}{3}} \text{Sin } \omega t \, dwt$$

$$= \frac{3V_m}{\pi} [- \text{Cos } \omega t]$$

$$= \frac{3V_m}{\pi} [-\text{Cos } (\alpha + \tfrac{2\pi}{3}) + \text{Cos } (\alpha + \tfrac{\pi}{3})]$$

$$V_a = \frac{3V_m}{\pi} \text{Cos } \alpha \text{ ----(1)}$$

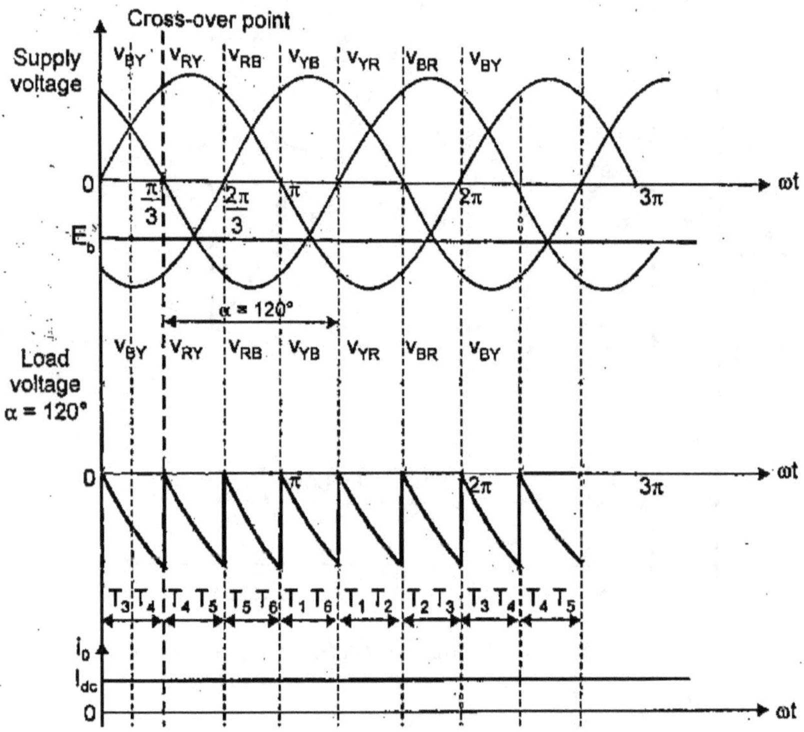

Fig.2.11. Output Waveform (α>900)

WKT,

$$V_a = E - I_a R_a \text{ ----(2)}$$

$$E = K\omega_m \text{ ----(3)}$$

$$T = KI_a \text{ ----(4)}$$

Equating (1) & (2)

$$E - I_a R_a = \frac{3V_m}{\pi} \cos \alpha \quad ----(5)$$

Subs (3) & (4) in (5)

$$K\omega_m - \frac{T}{K} R_a = \frac{3V_m}{\pi} \cos \alpha$$

$$K\omega_m = \frac{3V_m}{\pi} \cos \alpha + \frac{T}{K} R_a$$

$$\omega_m = \frac{3V_m}{K\pi} \cos \alpha + \frac{T}{K^2} R_a$$

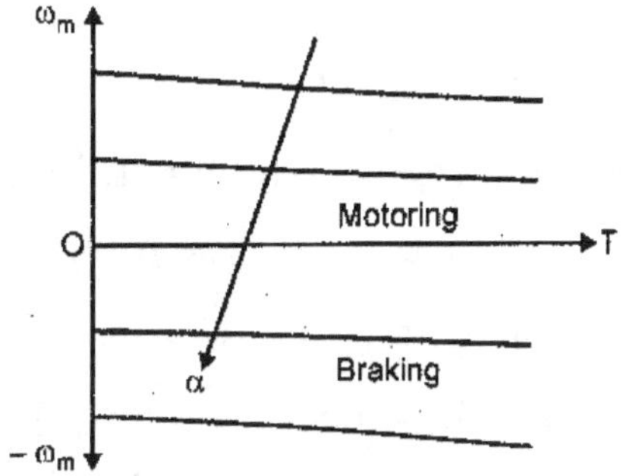

Fig.2.12. Torque Speed Characteristics

2.4. STEADY STATE ANALYSIS OF THREE PHASE HALF CONVERTER FED SEPARATELY EXCITED DC MOTOR DRIVE

- It is used upto 115KW applications.

- Here, average output voltage & current will be either positive.

- Three phase AC supply is connected to the input terminals of converter circuit and the armature terminals of the motor are connected with the output terminals of the converter circuit.

- This converter circuit consists of three thyristors and three diodes in it.

- Firing angle for the thyristors can be varied from 0 to α.

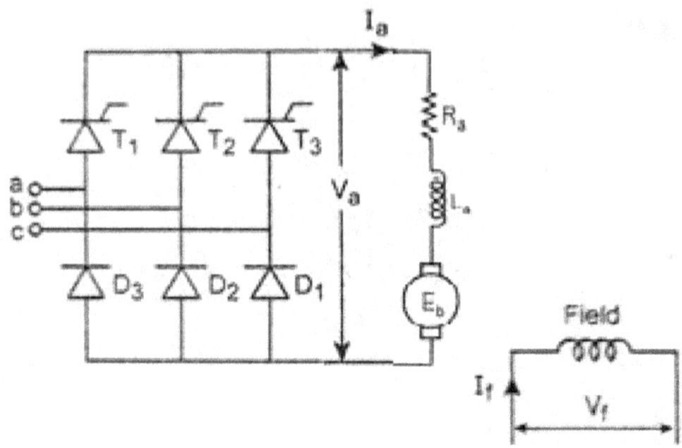

Fig.2.13. Three Phase Half converter fed DC Drive

- During the period $\frac{\pi}{6}$ to $\frac{7\pi}{6}$, T_1&D_1 conducts and the supply voltage V_{ac} gets applied across the motor terminals.

- From $\frac{7\pi}{6}$, V_{ac} starts to be negative, motor current starts to flow through freewheeling diode and T_1&D_1 gets turned OFF. So the armature voltage becomes 0.

$$V_a = \frac{3}{2\pi} \int_\alpha^\pi V_m \, \text{Sin} \, \omega t \, d\omega t$$

$$= \frac{3V_m}{2\pi} \int_\alpha^\pi \text{Sin} \, \omega t \, d\omega t$$

$$= \frac{3V_m}{2\pi} [-\text{Cos} \, \omega t]$$

$$= \frac{3V_m}{2\pi} [-\text{Cos} \, \pi + \text{Cos} \, \alpha]$$

$$= \frac{3V_m}{2\pi} (1+\text{Cos} \, \alpha) \, ----(1)$$

WKT,

$$V_a = E + I_a R_a \, ----(2)$$

$$E = K\omega_m \, ----(3)$$

$$T = K I_a \, ----(4)$$

Equating (1) & (2)

$$E + I_a R_a = \frac{3V_m}{2\pi}(1 + \cos\alpha) \text{----(5)}$$

Subs (3) & (4) in (5)

$$K\omega_m + R_a \frac{T}{K} = \frac{3V_m}{2\pi}(1 + \cos\alpha)$$

$$K\omega_m = \frac{3V_m}{2\pi}(1 + \cos\alpha) - R_a \frac{T}{K}$$

$$\omega_m = \frac{3V_m}{2\pi K}(1 + \cos\alpha) - R_a \frac{T}{K^2}$$

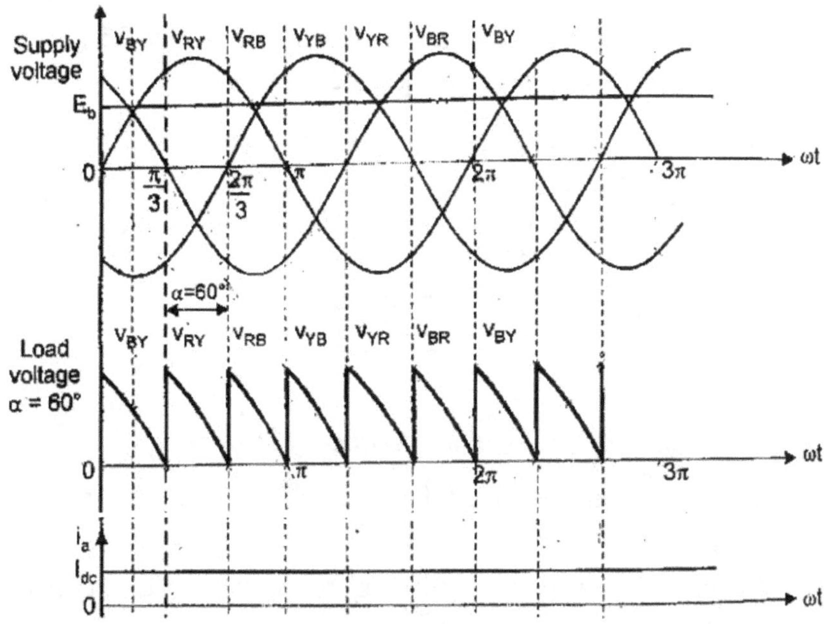

Fig.2.14. Output Waveform

2.5. CONTROL STRATERGIES - TIME RATIO & CURRENT LIMIT CONTROL

- Average output voltage of the converter can be controlled by varying the firing angle α, which turns ON & OFF the thyristors.

- The control strategies are, Time ratio control and current limit control.

a) Time ratio control

Here, in this control, the value of $\frac{T_{ON}}{T}$ is varied and this gives out two controlling systems.

1. *Constant Frequency System*

- Here, the T_{on} is varied & the chopping frequency is kept constant.

- So the width of the pulse gets varied and this type of control is known as Pulse Width Modulation (PWM).

- The below fig shows the waveform for $\alpha = 25\%$ (i.e) $T_{on} = \frac{1}{4}T$ & $\alpha = 75\%$ (i.e) $T_{on} = \frac{3}{4}T$

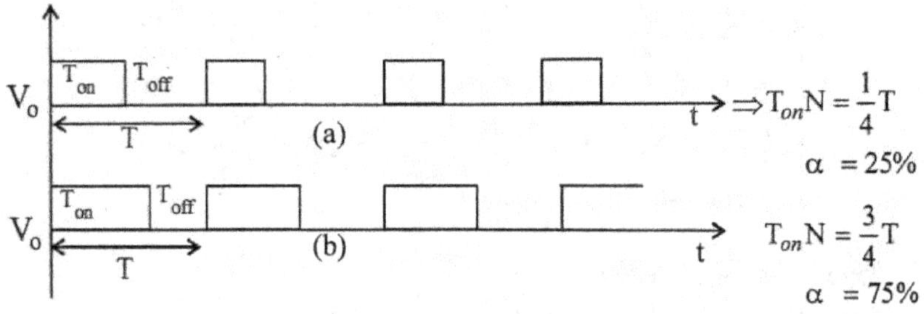

Fig.2.15. Constant frequency System

- α can be varied from 0 to infinity, which varies the output voltage V_o from 0 to V_s.

- Constant frequency control gives low ripple and it has faster response.

- This type of control is used for chopper drives.

2. *Variable Frequency System*

- Here, the T_{on} or T_{off} is kept constant and the chopping frequency is varied.

- This method of controlling the duty cycle is known as frequency modulation scheme.

- The fig.2.16. shows $T_{on} = \frac{1}{2} T$ (i.e) 50% & $T_{on} = \frac{3}{4} T$ (i.e) 75%

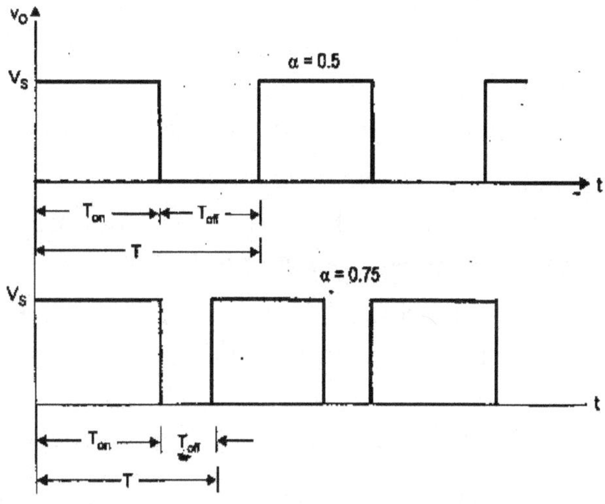

Fig.2.16. Variable frequency System (Ton Constant)

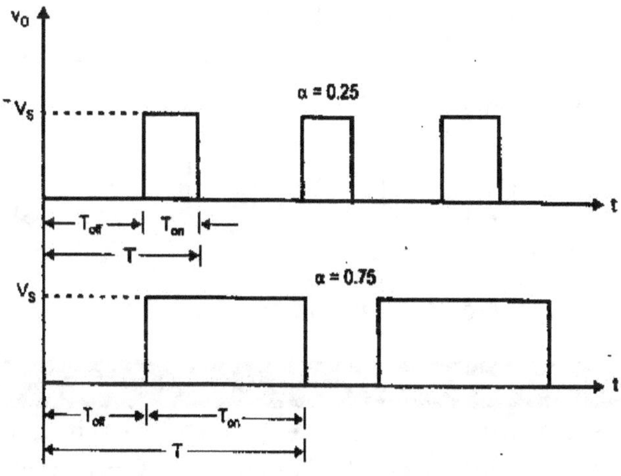

Fig.2.17. Variable frequency System (Toff Constant)

b) Current limit control

- In this type of control, the chopper is switched ON and OFF, so that the current is maintained between two limits (I_{omin} to I_{omax}).

- Now, the chopper is switched ON, the output current starts from I_{omin} and gradually gets increased.

- Once the output current reaches I_{omax}, the chopper is switched OFF, so the output current starts to decreases gradually to I_{omin}.

- Again the chopper is switched ON, once the load current reaches I_{omin}.

- Current limit control method is used only when the load has energy storage elements.

- Current limit control method is possible either in Constant Frequency and Variable Frequency.

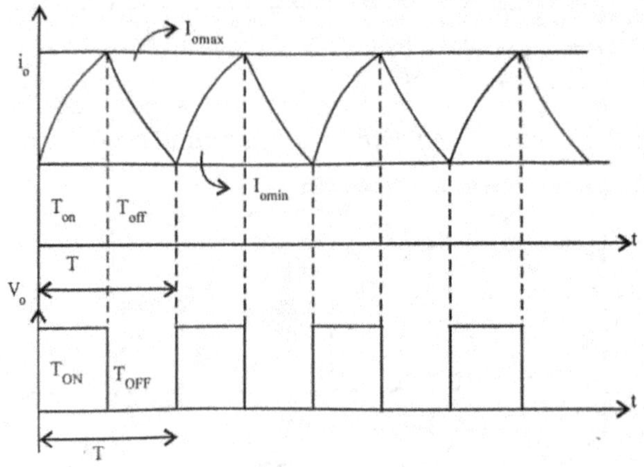

Fig.2.18. Current limit control

2.6. CHOPPER FED DC DRIVE

Chopper is one in which converts fixed DC to variable DC and it is also known as DC-DC converter. Choppers are made by using self-commutated devices like MOSFET, IGBT, GTO, power transistors etc.,

Adv: High efficiency, flexible, light weight, small size, quick response, etc.,

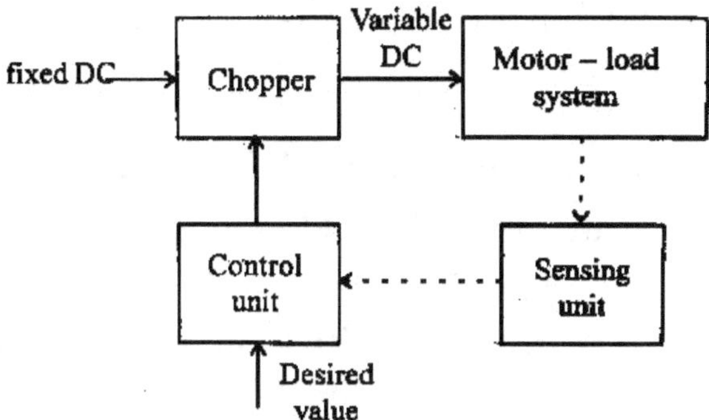

Fig.2.19. Chopper fed DC Drive

TYPES OF CHOPPERS

a) I Quadrant Chopper or Type A Chopper

- Operates in forward motoring mode and the flow of energy is from source to load.

- Output voltage and current is positive.

- Here, the output voltage Vo is less than the supply voltage Vs, so it is also known as step down chopper.

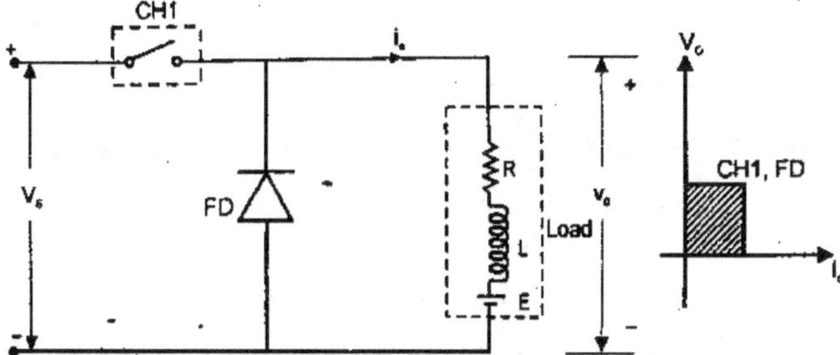

Fig.2.20. Type A Chopper & Quadrant of Operation

Operation

✓ **When CH₁-ON:** CH₁ - forward biased, FD-reverse biased. Inductance gets charged when is CH₁-ON. Load current increases due to the presence of inductance.

✓ **When CH₁-OFF:** CH₁ - reverse biased, FD – forward biased. Load is disconnected from the source. Output voltage Vo=0. Energy stored in inductance freewheels and decays slowly.

b) II Quadrant Chopper or Type B Chopper

- Operates in forward braking mode and the flow of energy is from load to source.

- Output voltage is positive and output current is negative.

- Here, the output voltage Vo is greater than the supply voltage Vs, so it is also known as step up chopper.

- It is also known as regenerative chopper.

- In this chopper, load must contain DC source E in it.

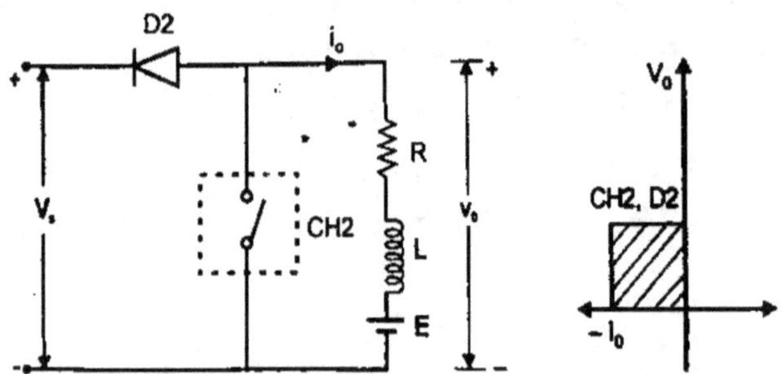

Fig.2.21. Type B Chopper & Quadrant of Operation

Operation

✓ **When CH₂-ON:** CH-Forward biased, D₂-Revese biased. So the output voltage Vo is 0. Circuit is driven by E in load through L and CH₂. Inductance gets charged when CH₂ is ON.

✓ **When CH₂-OFF:** CH₂-Reverse biased, D₂ - Forward biased. So power flows from source to load.

c) II Quadrant Type A Chopper or Type C Chopper

- Here, the circuit has 2 choppers CH₁ & CH₂, 2 diodes D₁ & D₂ with RLE load.

- Chopper circuit operates in I & II quadrants.

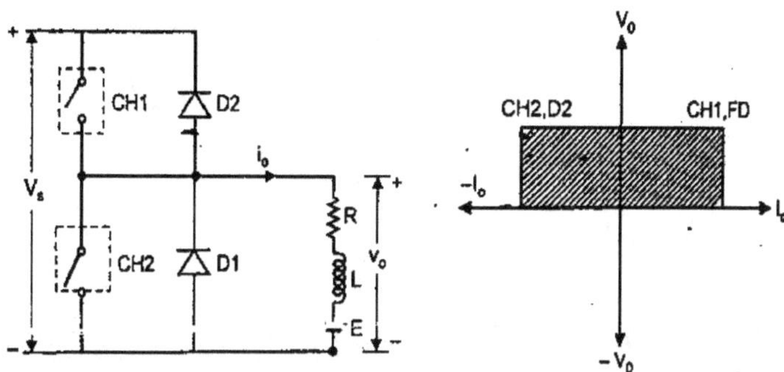

Fig.2.22. Type C Chopper & Quadrant of Operation

I Quadrant:

- CH_1 – ON, Current flows from source to load through $Vs(+)$-CH_1-R-L-E-$Vs(-)$.

- L gets charged.

- Voltage and current is always positive.

- Due to the energy in inductance, CH_1 – OFF, D_1 turns ON. Stored energy in the inductor starts to freewheels and decays slowly.

II Quadrant:

- CH_2 – ON, Current flows from load to source.

- Circuit is driven by E in load through CH_2 & L.

- Due to the energy in L, CH_2-OFF, D_2 turns ON.

- Path of current is $Vs(+)$-D_2-R-L-E-$Vs(-)$

- Voltage is positive and current is negative.

d) II Quadrant Type B Chopper or Type D Chopper

- Here, the circuit has 2 choppers CH_1 & CH_2, 2 diodes D_1 & D_2 with RLE load.

- Chopper circuit operates in I & IV quadrants.

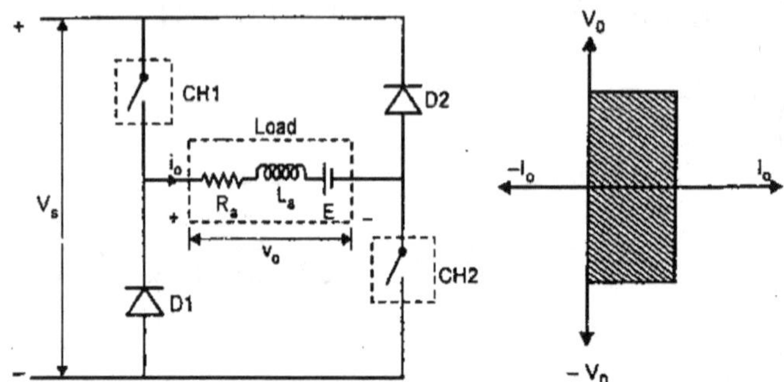

Fig.2.23. Type D Chopper & Quadrant of Operation

I Quadrant:

- CH$_1$ & CH$_2$ – ON, Current flows from source to load through Vs(+)-CH$_1$-RLE-CH$_2$-Vs(-).
- L gets charged.
- Due to the energy in L, CH$_1$-OFF, D$_1$ turns ON.
- So, the energy stored in L gets freewheels CH$_2$ & D$_1$.
- Both voltage and current is positive.

IV Quadrant:

- D$_1$ & D$_2$ – ON, current flows from load to source.
- Here, the voltage is negative and the current is positive.

e) Four Quadrant Chopper or Type E Chopper

Here, the circuit has 4 choppers CH$_1$, CH$_2$, CH$_3$ & CH$_4$, 4 antiparallel diodes D$_1$, D$_2$, D$_3$ & D$_4$ with RLE load.

I Quadrant:

- CH$_1$ & CH$_2$ – ON, current flows from source to load through Vs(+)-CH$_1$-R-L-E-CH$_2$-Vs(-).
- L charges
- Due to the energy stored in L, CH$_1$-OFF & D$_4$ – ON.
- Stored energy in L freewheels through D$_4$-R-L-E-CH$_2$.
- Voltage and the current will be positive.

- It is known as a forward motoring mode.

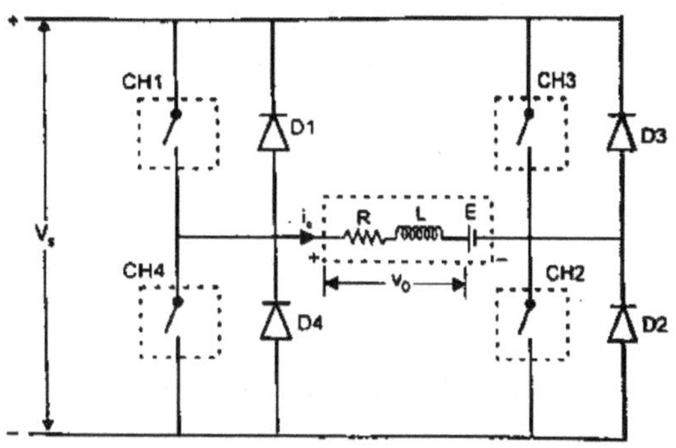

Fig.2.24. Type E Chopper & Quadrant of Operation

II Quadrant:

- CH_4 & D_2 – ON, Current flows from load to source. Current moves in reverse direction. So the current is negative.

- Energy from E of the load flows in the circuit through CH_4 & D_2.

- L gets charged.

- Due to the energy in L, CH_4 – OFF & D_1 – ON.

- So the current passes through $Vs(+)$-D_1-R-L-E-D_2-$Vs(-)$.

- Here, voltage is positive and the current is negative.

- It is known as forward braking or forward regeneration mode.

III Quadrant:

- CH_3 & CH_4 – ON, current flows from the source to load through $Vs(+)$-CH_3-R-L-E-CH_4-$Vs(-)$.

- L gets charged.

- Due to the energy in L, CH_3-OFF, D_2-ON.

- Energy stored in L freewheels between CH_4 & D_2.

- The polarity of E in load is reversed, so the voltage is negative. Current is also negative.
- It is known as Reverse motoring mode.

IV Quadrant:

- CH_2 & D_4 – ON, Current from E of load flows from load to source. Current moves in reverse direction. So the current is negative.
- Current flows through D_4-R-L-E-CH_2.
- L gets charged.
- Due to the energy in L, CH_2-OFF & D_3-ON.
- So the current passes through $Vs(+)$-D_3-E-L-R-D_4-$Vs(-)$.
- Here both voltage and current are negative.
- It is also known as reverse braking or reverse regeneration mode.

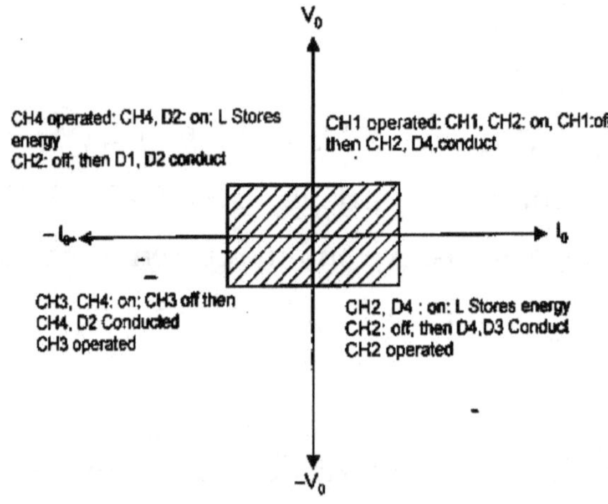

Fig.2.25. Type E Chopper Quadrant of Operation

SOLVED PROBLEMS

1. **A 200V, 875 RPM, 150A separately excited DC motor has an armature resistance of 0.06Ω. It is fed from a single phase fully controlled rectifier with an AC source of**

220V, 50 Hz. Assuming continuous condition, calculate (i) firing angle for rated motor torque and 750 RPM. (ii) Motor speed for $\alpha=160^0$ and rated torque.

Given:

Motor rating = 200V, N=875 RPM, Ia=150A,

Ra=0.06Ω, V=220V & f=50Hz.

Solution:

At rated operation,

Eb = V-I_aR_a = 200-(150-0.06) = 191 V

i) Firing angle for rated motor torque and 750 RPM

E_b at 750 RPM, $E_b = \dfrac{750}{875}$*191 = 163.7 V

$V_a = E_b + I_aR_a$ = 163.7+(150*0.06) = 172.7 V

Average output voltage of single phase fully controlled rectifier is,

$V_a = \dfrac{2V_m}{\pi}$ Cos α

$172.7 = \dfrac{2*\sqrt{2}*220}{\pi}$ Cos α

$\alpha = 29.3^0$

ii) Motor speed for α = 1600 and rated toque

$V_a = \dfrac{2V_m}{\pi}$ Cos α = $\dfrac{2*\sqrt{2}*220}{\pi}$ Cos 160^0 = -186.12V

$V_a = E_b + I_aR_a$

-186.12 = E_b+(150*0.06)

E_b = -195 V

Speed = $\dfrac{-195.12}{191}$*875

N = -893.8 RPM

2. A 230V, 960 rpm and 200A separately excited DC motor has an armature resistance of 0.02 Ω. The motor is fed

from a chopper which provides both motoring and braking operations. The source has a voltage of 230V. Assuming continuous conduction. (i) Calculate duty ratio of chopper for motoring operation at rated toque and 350 rpm. (ii) Calculate duty ratio of chopper for braking operation at rated torque and 350 rpm. (iii) If maximum duty ratio of chopper is limited to 0.95 and maximum permissible motor current is twice the rated, calculate maximum permissible motor speed obtainable without field weakening and power fed to the source (iv) If motor field is also controlled (v) Calculate field current as a fraction of its rated value for a speed of 1200 RPM.

Solution:

i) *Duty ratio of chopper for motoring operation at rated torque and 350 rpm*

At rated operation, E_b = 230-(200*0.02) = 226V

E_b at 350 rpm = $\dfrac{350}{960}$*226 = 82.4 V

Motor terminal voltage, $V_a = E_b + I_a R_a$

$\qquad$ = 82.4+(200*0.032) = 86.4V

Duty ratio $\delta = \dfrac{86.4}{230}$ = 0.376

$\qquad \delta = 0.376$

ii) *Duty ratio of chopper for braking operation at rated torque and 350 rpm*

Maximum available armature voltage, Va = Eb+IaRa

$\qquad$ = 82.4-(200*0.02) = 78.4V

Va = (1-δ)Va

78.4 = (1-δ) 230

δ = 0.6591

iii) Power fed back to the source

Maximum available armature voltage,

$$V_a = 0.95*230 = 218.5V$$

$$Ia = 2*200 = 400A \text{ (twice the rated current)}$$

$$E = V_a+I_aR_a = 218.5+(400*0.02) = 226.5 \text{ V}$$

Maximum permissible motor speed,

$$= \frac{960}{226}*226.5 = 962 \text{ rpm}$$

Assuming lossless chopper, power fed into the source,

$$= V_aI_a = 218.5 * 400 = 87.4 \text{ KW}$$

$$P = 87.4 \text{ KW}$$

iv) Field current as a fraction of its rated value for a speed of 1200 rpm

As in (iii) E=226.5 V for which at rated field current, speed = 960 rpm. Assuming linear magnetic circuit E_b will be inversely proportional to filed current.

Field current as a ratio of its rated value $= \dfrac{960}{1200} = 0.8.$

UNIT 3
INDUCTION MOTOR DRIVES

Speed Control of 1Φ and 3Φ induction motor can be controlled using solid state devices by following methods,

- Stator Voltage control method
- Stator frequency control method
- Rotor resistance control
- Slip power recovery control
- Pole changing method of control
- Eddy current coupling

3.1. STATOR VOLTAGE CONTROL

- Stator voltage control is used to control the speed of the induction motor by controlling the stator voltage.

- Here the supply frequency is kept constant.

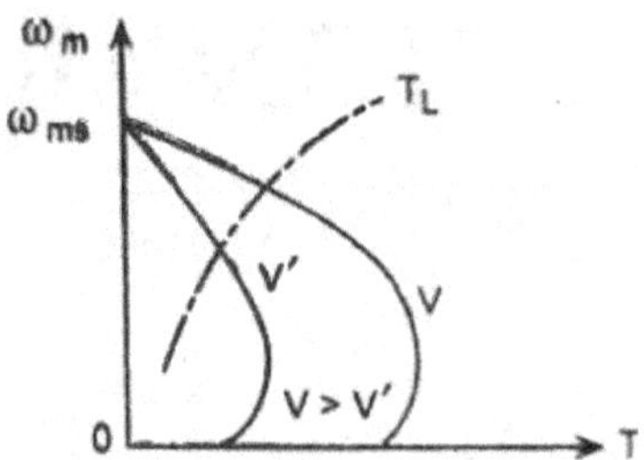

Fig.3.1. Stator Voltage Control

- Torque equation of the induction motor is,

$$T = \frac{3}{\omega_s} \left[\frac{V^2}{\left(R_S + \frac{R_r^l}{S}\right)^2 + (X_S + X_r^l)^2} \right] \frac{R_r^l}{S}$$

From the above equation, we can see, torque is proportional to the square of stator voltage.

- So, as the voltage is reduced the torque gets reduced double the times.

- Stator voltage control is an excellent method for reducing the starting current with higher efficiency for light loads.

- This method of control is used to control the induction machine operating below the rated speed.

- Here the speed control is done by using the AC voltage controllers (thyristor switches connected between the supply and motor).

- A three phase full wave AC voltage controller consists of 6 thyristors, 2 thyristors connected in antiparallel manner.

- A three phase half wave AC voltage controller consists of 3 thyristors and 3 diodes.

- Similar to three phase, there is also single phase AC voltage controller. Full controller has 2 thyristors connected in antiparallel manner and half controller has 1 thyristor and 1 diode.

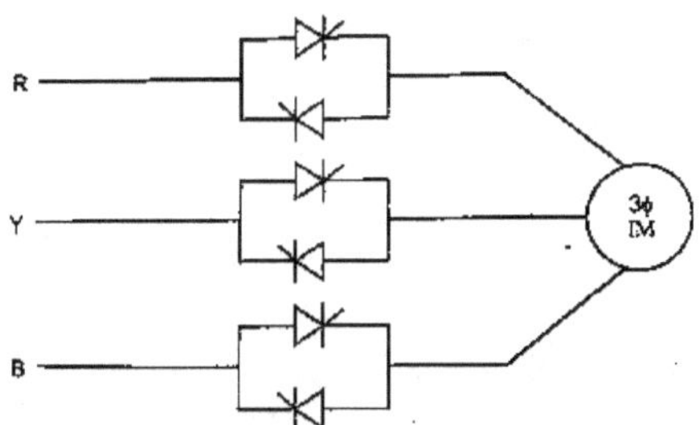

Fig. 3.2. 3Φ Full Wave AC Voltage Controller

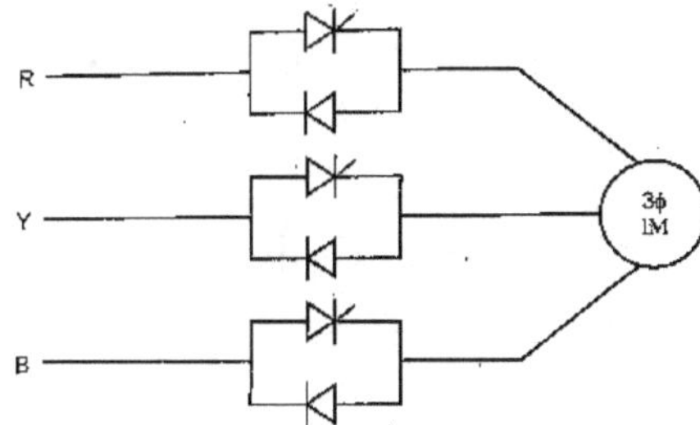

Fig. 3.3. 3Φ Half Wave AC Voltage Controller

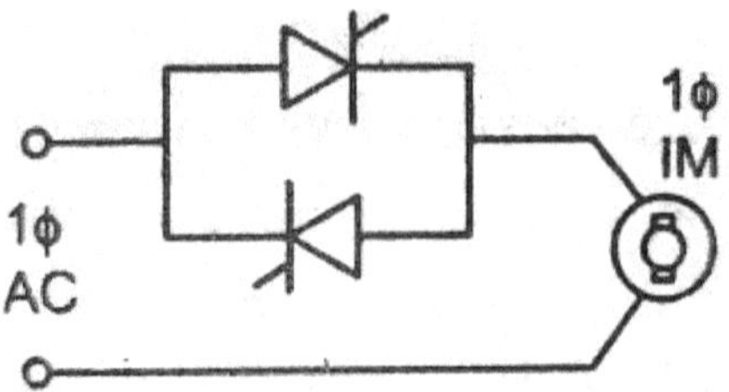

Fig.3.4. 1Φ Thyristor Voltage Controller

- There are two types of control using AC voltage controller,

a) On-off control (or) integral cycle control

- Here the thyristors are used as switches.

- Thyristors connects the source and load for few cycles and disconnects for few cycles.

- This method is also known as integral cycle control.

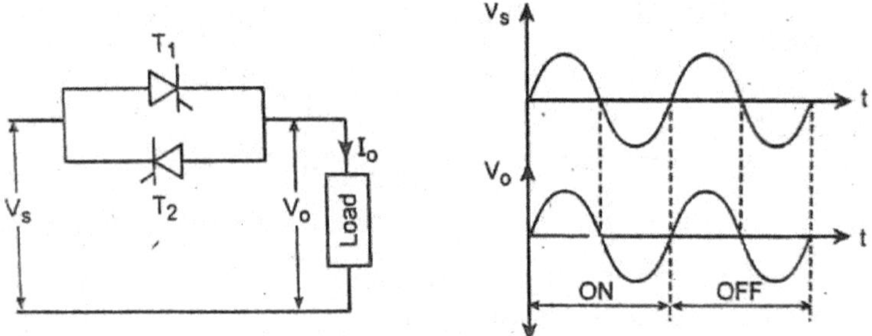

Fig.3.5. On Off Control

b) Phase angle control

- Here the power flow from the source to load is controlled by varying the firing angle the thyristor.

Advantages:

- ✓ Control circuit is very simple
- ✓ Compact and less weight
- ✓ High response time
- ✓ More economic

Disadvantages:

- ✓ Input power factor is low
- ✓ Operating efficiency is low as losses are high.
- ✓ Voltage and current waveforms are highly distorted due to harmonics.

Applications:

- ✓ Low power applications like fan, blowers, centrifugal pumps, etc.,

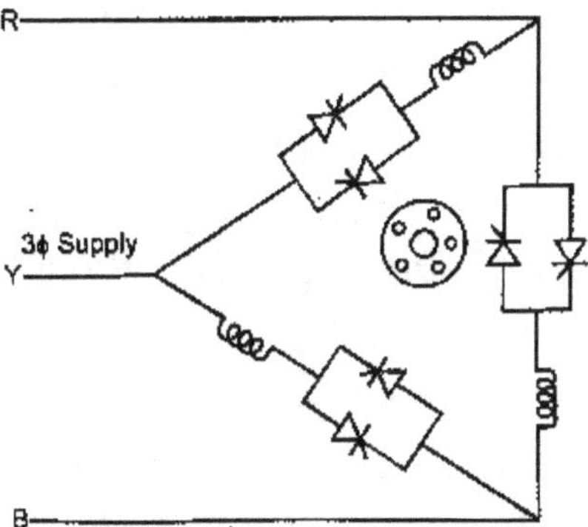

Fig.3.6. 3Φ Full Wave AC Voltage Controller for delta connected load

3.2. VOLTAGE/ FREQUENCY CONTROL

- Most commonly used method of speed control.

- Here, increase in supply frequency increases the speed and reduces the maximum torque of the motor. But increase in voltage increases the maximum torque of the motor.

  f increases: N increases: T_{max} decreases

  V increases: T_{max} increases

- Change in voltage and frequency is the powerful method for speed control of induction motor.

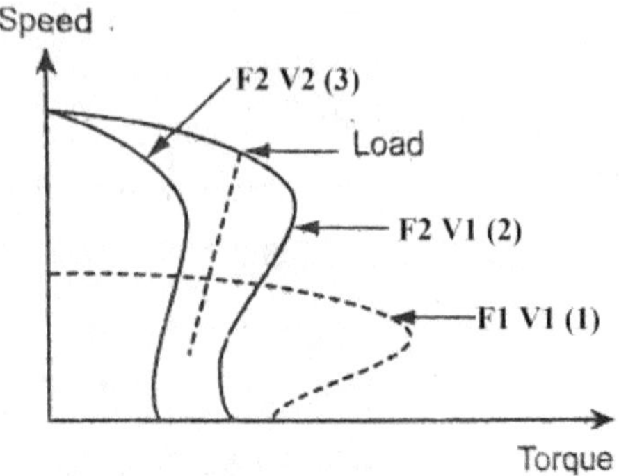

Fig.3.7. Torque Speed Curve

- From the torque speed characteristics, V_1 and F_1 are the reference voltage and frequency – Curve 1.

- Now the frequency is increased to F_2 but voltage remains unchanged. So the speed increases and the maximum torque decreases. During load, there no steady state operation and motor stalls – Curve 2.

- Now the voltage is increased to V_2 and frequency remains unchanged (i.e) at F_2 itself. So maximum torque increases and the motor attains the new state. Steady state operation is achieved – Curve 3.

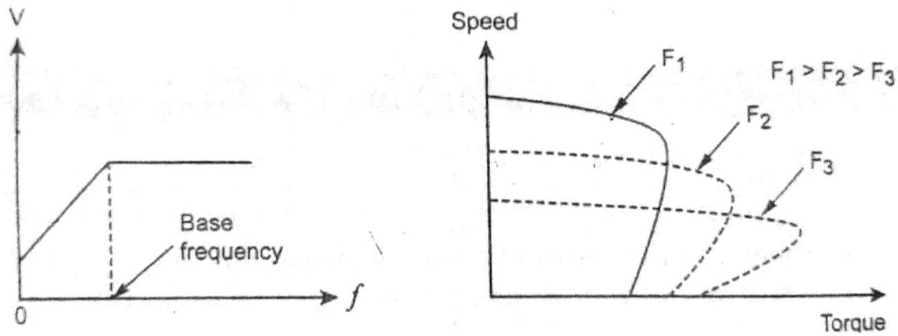

Fig.3.8. Torque Speed Curves of V/F Control

- Main feature of V/F control method are,
    - ✓ Maximum torque should be constant.
    - ✓ Starting current is also constant.

Open Loop Control

- This control scheme consists of controlled rectifier (converts AC to DC), filter capacitor (remove the harmonics and ripples produced during the AC to DC conversion), Variable Voltage Variable frequency Inverter (converts DC to AC) and the induction motor whose speed is to be controlled.

- Frequency command f_{sn}^* is given to the inverter.

- Offset voltage Von is added and it multiplied by 2.22 to obtain the DC link voltage.

- Here the speed can't be controlled precisely and the slip cannot be maintained.

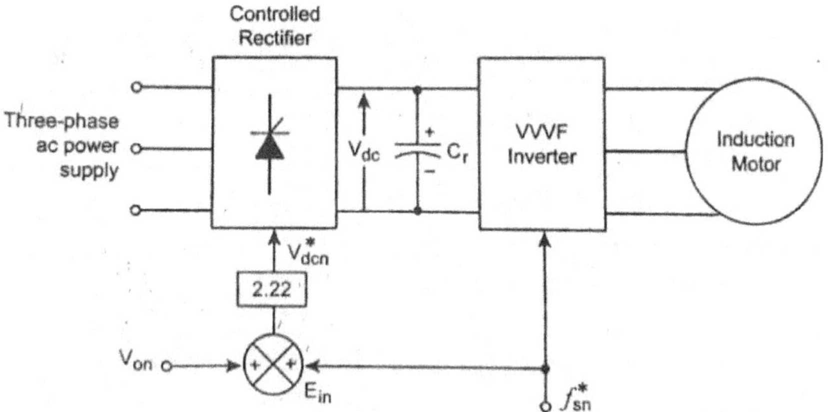

Fig.3.9. Open Loop induction motor drive with V/F Control Strategy

Closed Loop Control

- Actual rotor speed of the induction motor is sensed by tachogenerator and compared with reference speed value ω_r^*, thus produces error signal.

- Error signal is processed in PI controller and limiter to generate slip speed command ω_{sl}^*.

- Limiter is used to maintain the slip speed within the allowable speed.

- Slip speed command ω_{sl}^* is added with electrical rotor speed ω_r to obtain stator frequency command f_s^*.

- Stator frequency command f_s^*. is fed to the inverter and the rectifier.

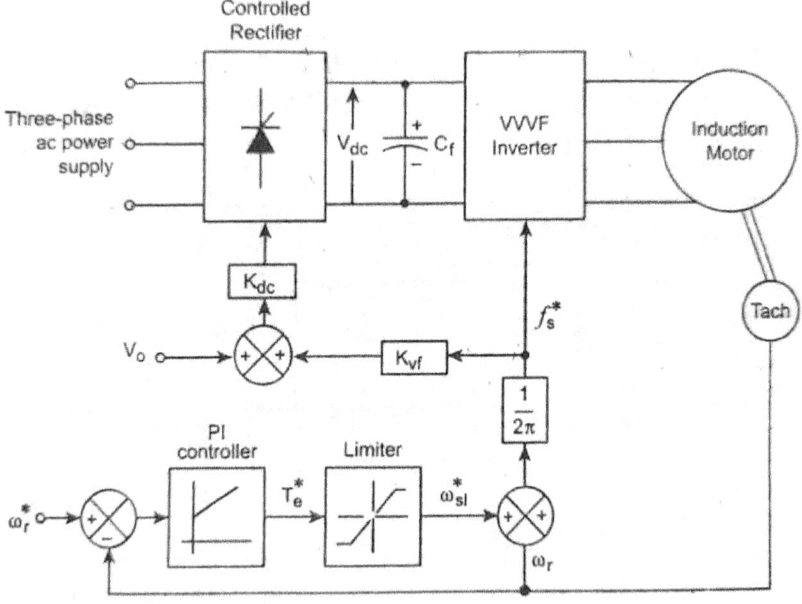

Fig.3.10. Closed Loop induction motor drive with V/F Control Strategy

- Before applying the frequency command f_s^* to the rectifier, f_s^* is added with offset voltage V_o and K_{dc}.

- K_{dc} is the constant proportionality between DC voltage and stator frequency.

- According to the signal received to the rectifier and the inverter, the speed of the induction motor is controlled as per the necessity.

3.3. CONSTANT AIR GAP FLUX CONTROL

- Motor speed can also be controlled by varying the supply frequency.

- Voltage induced in the stator is directly proportional to the product of supply frequency and air gap flux.

- Any reduction in the supply frequency, without change in terminal voltage, causes an increase in the airgap flux.

- Based on the design and magnetic material used in induction motor, the increase in flux saturate the motor.

- This saturation will increase the magnetizing current which distorts the line current and voltage, increases the core loss and stator copper loss, produce acoustic noise.

- When there is a increase in flux beyond the rated value causes saturation effects and when there is an decrease in flux causes lower torque capability of the motor.

- To have the efficient control, V/F ratio constant is maintained at rated value.

- Relation between air gap flux and torque is,

$$T_e = 3\left(\frac{P}{2}\right)\frac{1}{R_r}\,\emptyset_m^2\,\omega_{SL}$$

Te – Developed torque (NM)

P – Number of poles

Rr – Rotor resistance

Φm – Air gap flux linkage

ω_{SL} – Slip frequency

ω_{SL}^* – Slip speed reference

i_s^* – Stator current command

f_s^* – Stator frequency command

I_b – Base Current

I_C – Core loss current

ωr – Rotor Speed

ω_r^* - Speed reference

T_e^*- Torque reference

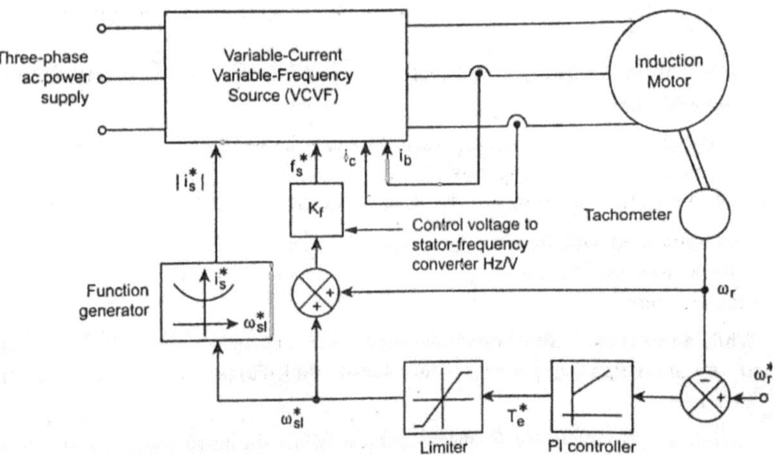

Fig.3.11. Constant air gap flux control

3.4. VARIABLE FREQUENCY INDUCTION MOTOR DRIVES

- Speed of the induction motor can also be converted by varying the supply frequency.

- This method of speed control is mainly applied in squirrel cage induction motor.

- Variable frequency can be obtained by,

  - (i)   Voltage Source Inverter (VSI)

  - (ii) Current Source Inverter (CSI)

  - (iii) Cyclo converter

(i) VOLTAGE SOURCE INVERTER (VSI)

- An inverter which converts DC Voltage to AC voltage and acts as a voltage source is known as Voltage Source Inverter (VSI).

- VSI is capable of supplying variable frequency variable voltage for speed control of induction motor.

- VSI is operated as stepped wave inverter or pulse width modulated inverter.

- If, VSI is to be operated as a stepped wave inverter, thyristors should be switched in sequence with time difference of T/6.

- Each thyristors are kept ON for the duration of T/2, where T is the time period of one cycle.

- Frequency of the inverter operation is varied by varying the time period T and the output voltage is varied by varying the input voltage.

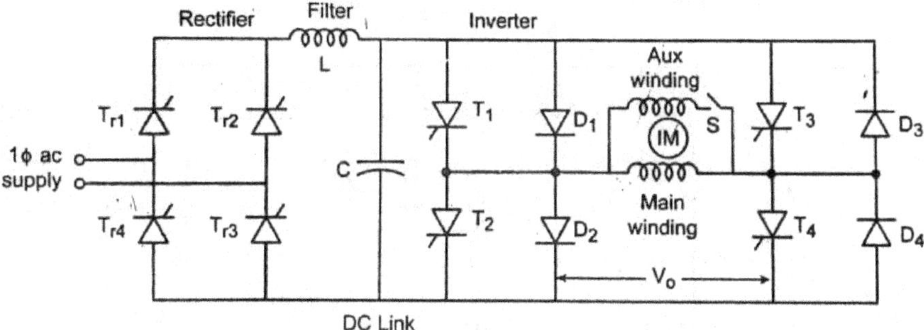

Fig.3.12. 1Φ VSI fed Induction Motor

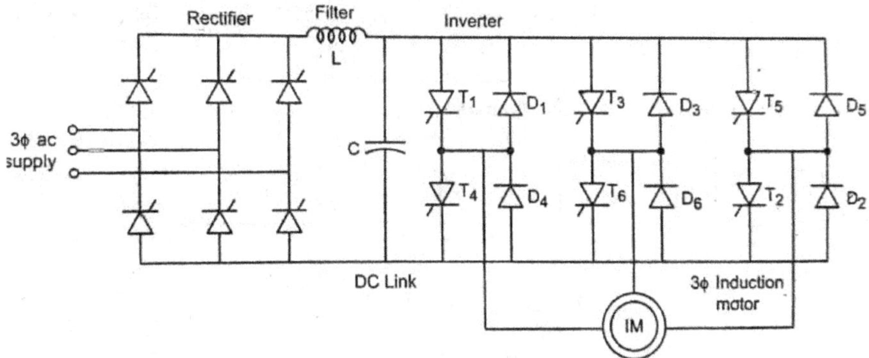

Fig.3.13. 3Φ VSI fed Induction Motor

- Various schemes of VSI are,

    a) ***Controlled rectifier based VSI*** – Consists of 3Φ controlled rectifier, filter and inverter. 3Φ supply fed to the controlled rectifier, it converts the 3Φ AC voltage into variable DC voltage and fed into the filter circuit which removes the ripples produced during the DC conversion. The output of filter circuit is fed into the inverter. The inverter produce variable frequency, variable voltage to control the induction motor connected to it.

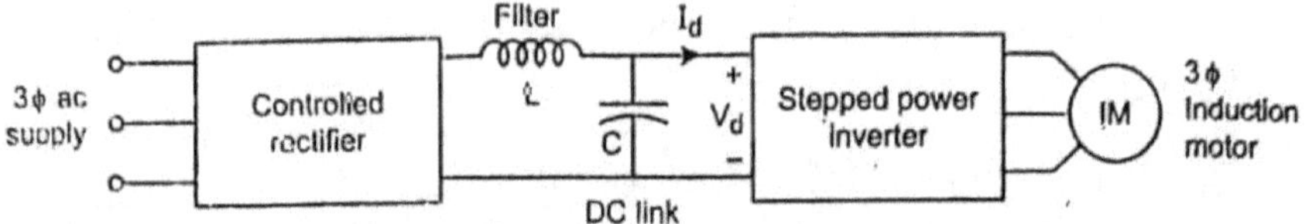

Fig.3.14. Controlled rectifier based VSI

    b) ***Dual Converter based VSI*** – Consists of dual converter, filter and inverter. If regeneration is process is done, the controlled rectifier is replaced by the dual converter. Filter is used to remove the harmonics produced during the DC voltage conversion process. Inverter is used to produce variable frequency and variable voltage to control the Induction motor connected to it.

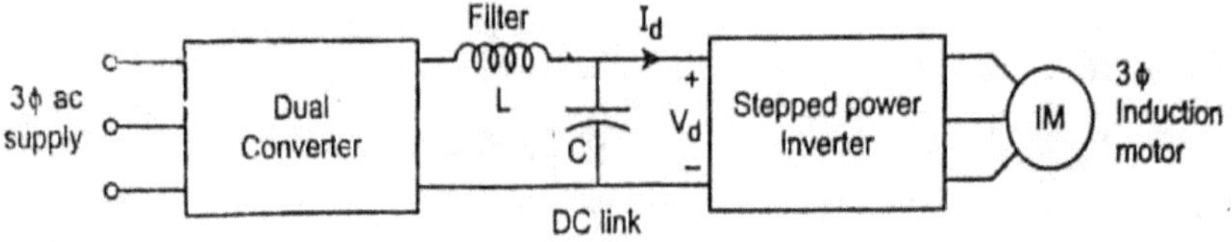

Fig.3.15. Dual Converter based VSI

    c) ***Diode bridge rectifier based VSI*** – Consists of diode rectifier, filter and the PWM inverter. Diode rectifier converts the 3Φ AC voltage into fixed DC voltage. Filter is used to remove the harmonics produced during the DC voltage conversion process. Fixed DC is applied to the PWM inverter. Here PWM techniques are applied to control the Induction Motor.

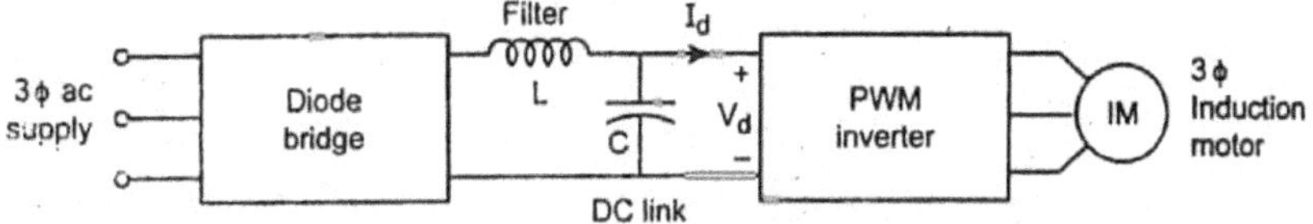

Fig.3.16. Diode bridge rectifier based VSI

d) Chopper based VSI – Consists of Chopper, filter and inverter. Chopper is used to convert fixed DC into variable DC. This type of VSI is used in system which provides high frequency output.

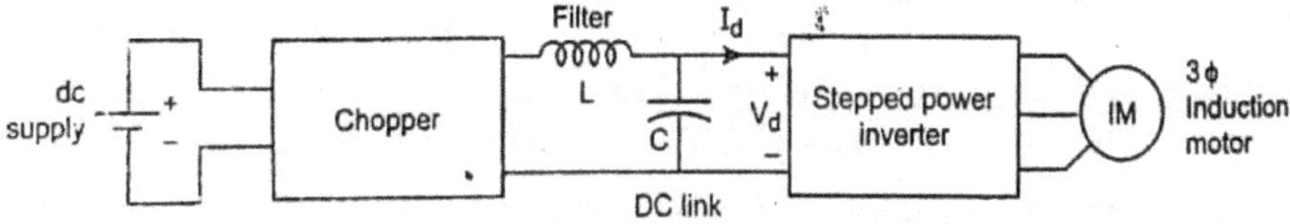

Fig.3.17. Chopper based VSI

e) Direct DC supply VSI – DC supply is directly connected to the PWM inverter. Here PWM techniques are applied to control the Induction Motor.

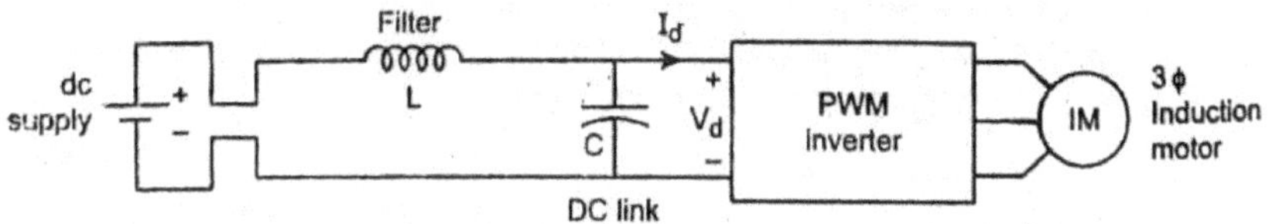

Fig.3.18. Direct DC Supply VSI

Disadvantages of VSI

✓ Larger low frequency harmonics

✓ Derating of motor

✓ Motor produces pulsating torque

(ii) CURRENT SOURCE INVERTER (CSI)

- In a DC link converter, if the DC link current is controlled, the inverter is called current source inverter.

- Current in the DC link is kept constant by impedance and capacitance of the filter.

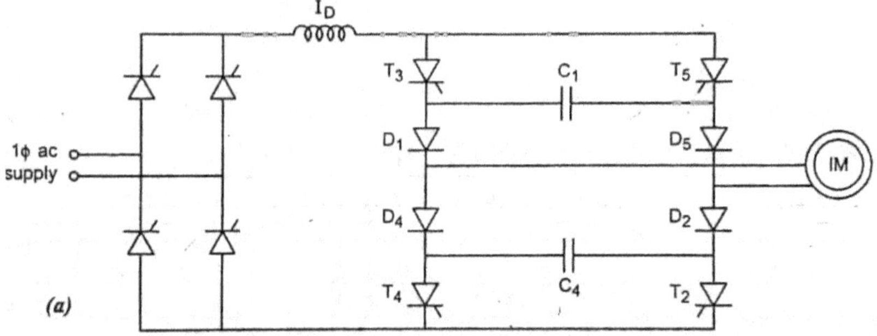

Fig.3.19. *1 Φ Current Source Inverter*

- CSI don't have any feedback diodes in it.

- CSI circuit consists of phase controlled rectifier, DC link and the inverter circuit.

- Here the DC link contains only the inductance, regeneration is possible by changing the polarity of voltages and maintain the current directions.

- There will be a stability problem at light load in CSI.

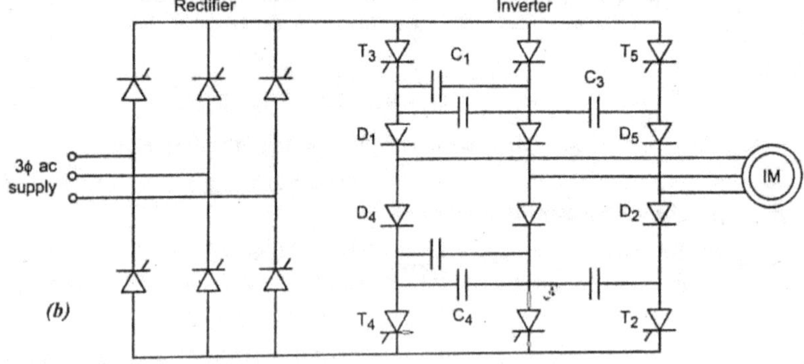

Fig.3.20. *3 Φ Current Source Inverter*

- Various schemes of CSI are,

a) Bridge rectifier based CSI fed drives

- Consists of 3Φ controlled bridge rectifier, inductor L_d and inverter.

- 3Φ AC voltage is converted into variable DC by using controlled bridge rectifier.

- DC voltage is converted into DC current source by passing through large inductor in series.

- The inverter frequency is controlled by varying the firing angle of the thyristor.

- The output of the inverter would be variable current variable frequency, which is used to control the speed of the motor.

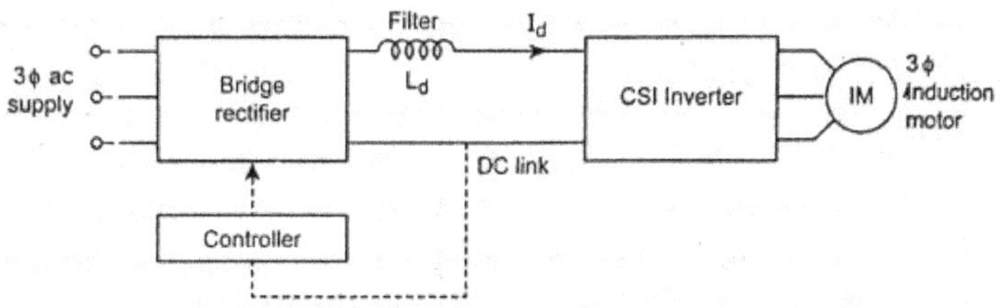

Fig.3.21. Bridge rectifier based CSI fed drives

b) Chopper based CSI fed drives

- Consists of three phase rectifier, DC chopper, inductor L_d and inverter.

- 3Φ rectifier is used to convert fixed AC to fixed DC voltage and it fed to the DC chopper.

- DC chopper converts fixed DC into variable DC voltage and fed to the inductor.

- Inductor is used to convert DC voltage into DC current and the current is fed to the inverter circuit.

- The inverter is variable current and variable frequency which is used to control the speed of the induction motor.

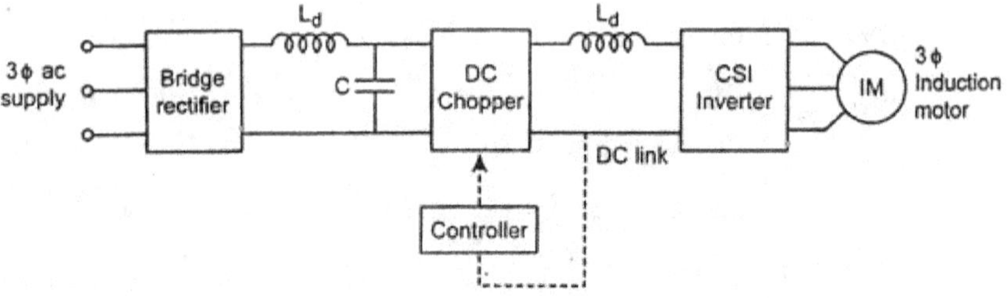

Fig.3.22. Chopper based CSI fed drives

COMPARISON OF CSI & VSI FED DRIVES

S. No.	CSI fed drives	VSI fed drives
1.	More reliable	Less reliable
2.	Does not require feedback diodes	Require feedback diodes
3.	Lower speed range	Higher speed range
4.	CSI drive is not suitable for multi motor drives	VSI drive is suitable for multi motor drives
5.	High cost and weight	Low cost and weight

(iii) CYCLOCONVERTER FED INDUCTION MOTOR

- Cycloconverter is used vary the frequency to control the speed of Induction motor.

- Cycloconverter is one in which converts AC line frequency into variable frequency.

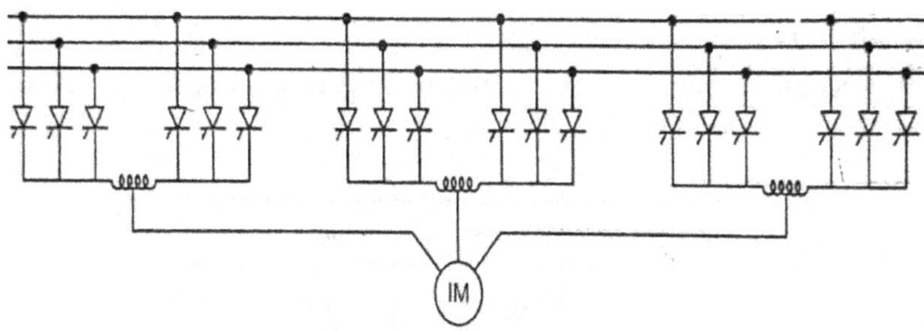

Fig.3.23. Cycloconverter fed Induction motor

- It operates by means of line commutation.

- Cost of the cycloconverter is high and it is complex control.

- It provides a very smooth low speed operation and has very low ripple torques.

- Output frequency of cycloconverter is limited to 1/3 times of the input frequency.

- Cycloconverter is capable of power transfer in either direction (i.e) between source and load.

3.5. ROTOR RESISTANCE CONTROL

- Stator side control of induction machine is applicable for both squirrel cage and slip ring induction motors.

- But the control through rotor side is applicable for only slip ring induction motor.

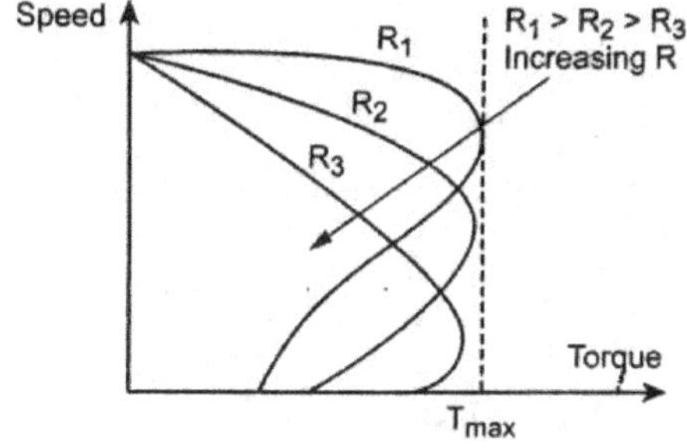

Fig.3.24. Speed torque characteristics of rotor resistance control

- While maximum torque is independent of rotor resistance, speed at which the maximum torque is produced changes with rotor resistance.

- Rotor power is rectified in a diode bridge rectifier and fed to the inductor to eliminate the ripples produced.

- Here the circuit consists of parallel connected resistance and chopper after the filter.

- This parallel connected setup is used to control the speed of the induction machine.

- Here the chopper is turned ON and OFF by the control circuits.

- When the chopper is ON, the current passes through the chopper and when the chopper is OFF, the resistance is included in the circuit.

- By varying the resistance in the circuit, speed of the machine varies.

- The control signal provided to chopper is by sensing the speed and current of the machine.

- The actual speed of the machine is sensed by the tacho generator, compared with the reference speed in comparator and generates an error signal.

- The error signal is amplified by speed amplifier and then sets the desired current reference.

- Now the reference current and the actual current is compared in the comparator and generates an error signal.

- The error signal is again amplified in current amplifier.

- This current signal controls the turning on and turning off of the chopper.

- By controlling the ON and OFF times of chopper, the resistance is determined and the speed of the machine is controlled.

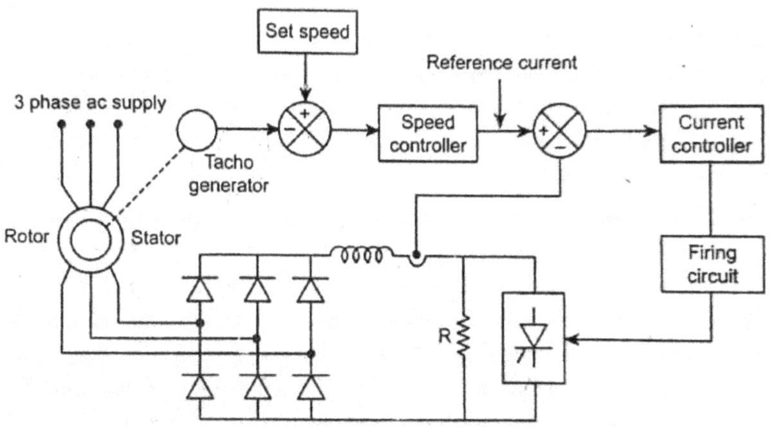

Fig.3.25. Closed loop rotor resistance control

3.6. QUALITATIVE TREATMENT OF SLIP POWER RECOVERY SCHEMES (ENERGY EFFICIENT DRIVE)

KRAMER SYSTEM

- Kramer system is applicable only for sub synchronous speed operation.

- Usually, slip power is wasted in the rotor circuit. Instead of wasting the rotor power, it is converted into useful power and fed back to the line. This method is known as Kramer system.

- System has 2 bridges, 1st bridge operates as a rectifier and the 2nd bridge operates as a inverter.

- Rotor power is taken from the rotor and given to the uncontrolled rectifier.

- So that the AC voltage is converted into DC voltage.

- Inductor L_d is used to remove the harmonics generated during the voltage conversion.

- After removing the ripples, the DC voltage is converted into AC voltage by using Inverter.

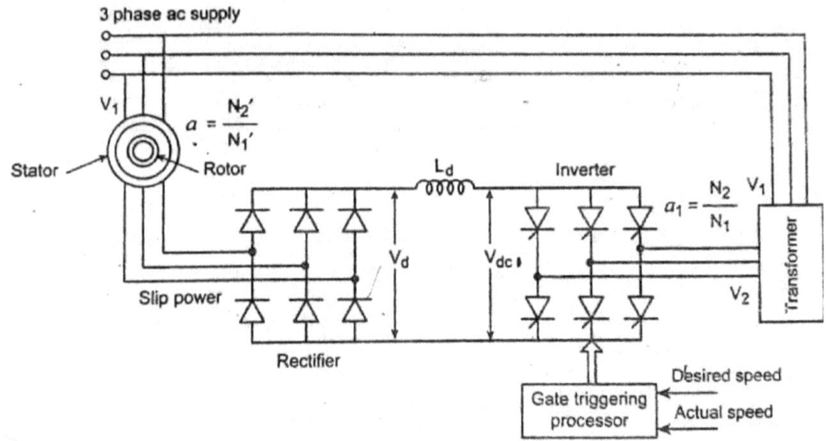

Fig.3.26. Static Krammer System

- The thyristors in the inverters are controlled by gate triggering processor.

- Now the AC voltage is stepped up using step up transformer and fed into the lines.

- Here slip power flows from rotor circuit to the supply; this method is also called as constant torque drive.

- The modified Kramer system is shown below,

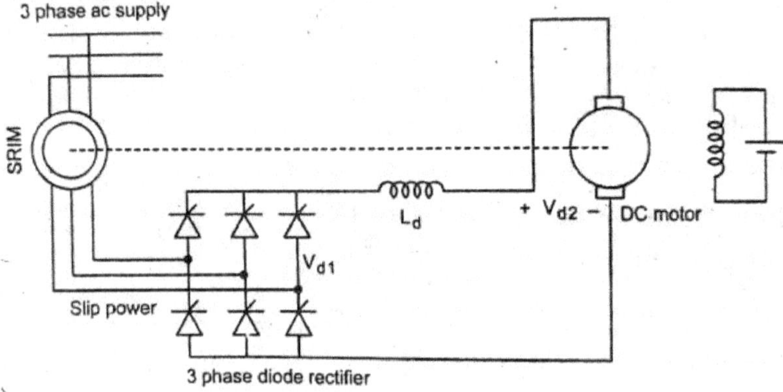

Fig.3.27. Modified Krammer System

- Here the rectified DC voltage is given to the inductor to remove the ripples in it and the DC power is given to the DC motor to operate it.

- Kramer system is most popular in large power pumps and fan type drives.

SCHERBIUS SYSTEM

- Scherius system is similar to that of the Kramer system. Only the diode bridge is replaced by the thyristor bridge.

- Consists of 2 controlled bridge circuits, filters and the transformer.

- Scherbius system is applicable for both sub synchronous and super synchronous speed operation.

Sub Synchronous Operation

- Slip power is removed from the rotor and it is fed into the AC supply system.

- During the operation, bridge 1 operates in rectifier mode and bridge 2 operates in inverter mode.

- Slip power from the rotor is fed to the bridge 1 (rectifier) which converts the AC voltage into DC voltage and fed to the inductor.

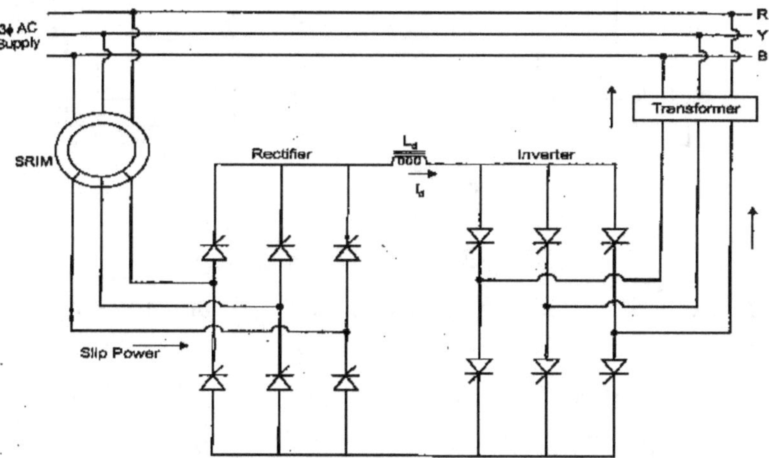

Fig.3.28. Static Scherbius System during Sub synchronous operation

- Inductor L_d is used to remove the harmonics generated during the voltage conversion.

- After removing the ripples, the DC voltage is converted in AC voltage by using Inverter.

- The thyristors in the inverters are controlled by gate triggering processor.

- Now the AC voltage is stepped up using step up transformer and fed into the lines.

Super Synchronous Operation

- Here additional power is applied to the rotor at slip frequency.

- During this operation, bridge 1 operates as Inverter and bridge 2 operates as Rectifier.

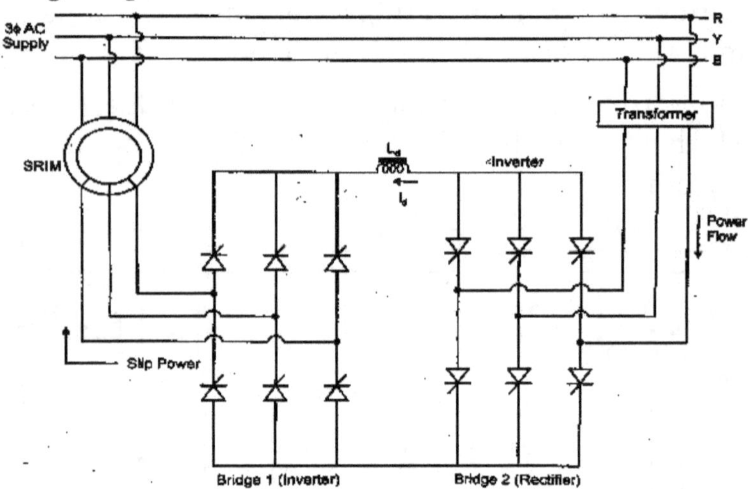

Fig.3.29. Static Scherbius System during Super synchronous operation

- The operation of bridge circuit can be changed by changing the firing angle. If the firing angle is less than 90^0 bridge circuit operates as rectifier and if the firing angle is more than 90^0 bridge circuit operates as inverter.

- Near synchronous speed, the rotor voltage is low, so in order to increase the voltage this system is used.

- Supply from 3Φ line taken and stepped down using the transformer and fed into the bridge 2.

- Bridge 2 operates as rectifier here, so that it converts AC voltage into DC voltage and fed to the inductor.

- Inductor L_d is used to remove the harmonics generated during the voltage conversion and fed into bridge 1.

- Bridge 1 here acts as a inverter, so that it converts the DC voltage into AC voltage and feds into the rotor.

3.7. CLOSED LOOP CONTROL

SINGLE QUADRANT CLOSED LOOP CONTROL

- The speed (ω_m) of the induction motor is sensed by tachogenerator, compared in the comparator with reference speed (ω_{ref}) and generates error signal.

- Error signal is fed to the speed controller and then fed into current limiter.

- Current limiter gives out the reference current (I_1^*).

- The actual motor current (I_1) is sensed by the current sensor and it is compared with reference current (I_1^*) and generates error signal.

- Then the error signal is fed to the current controller and then to the firing circuit.

- Firing circuit generates gating pulses for the thyristors in AC voltage controller.

- By varying the gating pulse, the speed of the induction motor can be controlled.

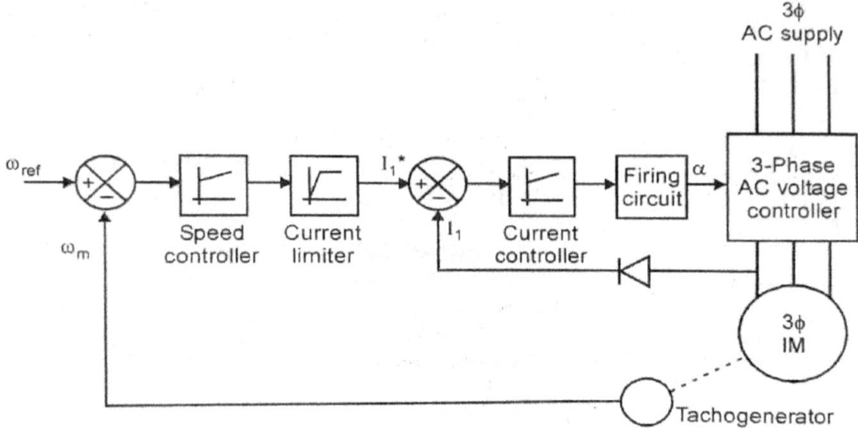

Fig.3.30. Single Quadrant Closed loop Control

FOUR QUADRANT CLOSED LOOP CONTROL

- This type of control consists of voltage and current controllers, an absolute value circuit block and master controller block.

- Absolute value circuit gives only the positive output voltage, whether input signal may be positive or negative.

- Master controller has 3 input signal (direction of rotation of the rotor $e_{\omega m}$, output signal of the firing circuit and stator current signal) and 1 output signal.

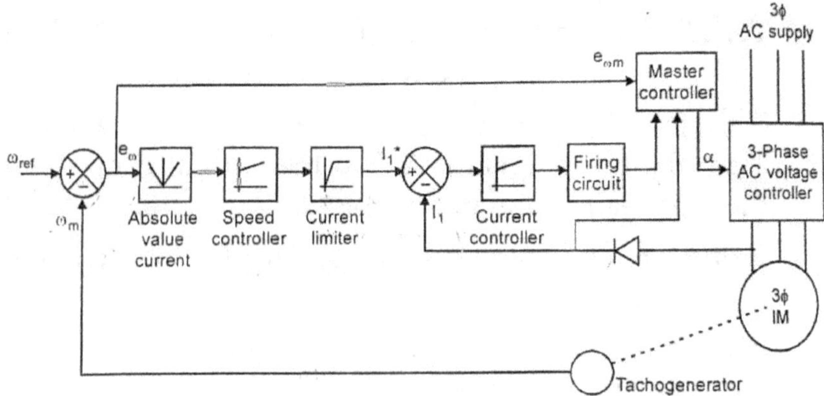

Fig.3.31. Four Quadrant Closed loop Control

- Master controller receives and processes these three signals and gives out the firing pulses to the thyristor sets RYB to have rotation in positive direction.

- To have reverse directions of rotation, the firing pulse can be $R^lY^lB^l$.

- Drive first accelerates and deaccelerates and finally settles at desired speed.

3.8. VECTOR CONTROL

- In vector control, the induction motor operates as a separately excited DC motor with high dynamic performance.

- In induction motor, there are no independent variables of armature current and field current.

- In DC machine, armature current and field current variables are independent.

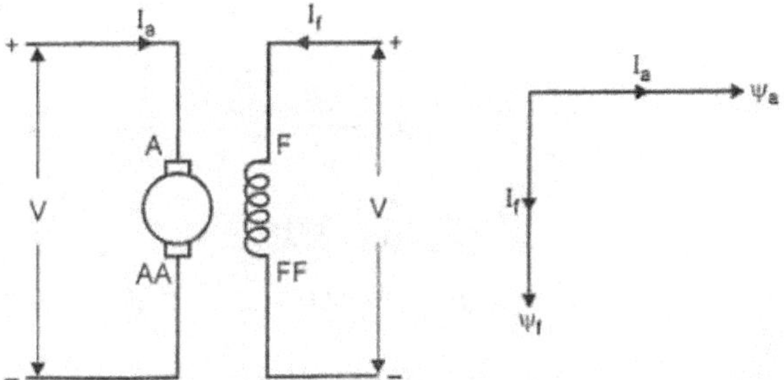

Fig.3.32. Separately excited DC motor

- In a DC machine, the developed torque is given by,

$$T \alpha \Phi_a$$

$$T = K\Phi I_a$$

$$T = KI_f I_a$$

K – Torque constant, I_a – Armature Current and I_f – Field Current

- Here, the field current is the direct axis component and armature current is the quadrature axis component.

  So, $T = KI_dI_q$

- In the above equation, both the components are independent.

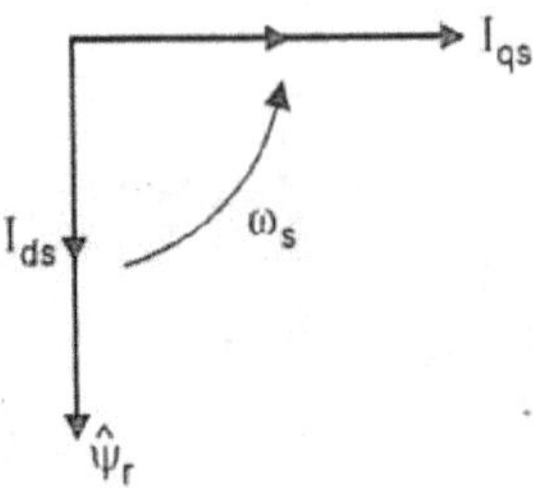

Fig.3.33. Space vector diagram of vector control

- This characteristics is found in the induction machine when applying vector control method.

- Here, the field flux linkage Ψ_f produced by I_f (I_d) is perpendicular to the armature flux linkage Ψ_a produced by I_a (I_q).

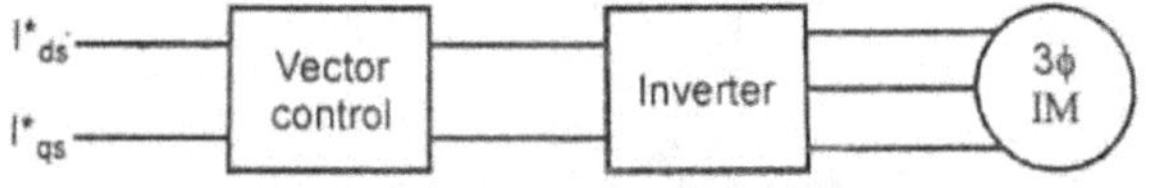

Fig. 3.34. Basic Block diagram of vector control

- Here there are two parts (control and machine part) which are separated by inverter unity gain.

- i^*_{ds} and i^*_{qs} are the reference control signal from microprocessor or microcontroller or DSP.

- First block converts the model into direct axis and quadrature axis stator current.

- Then it is fed to two phase to three phase (a,b,c) transform block and it is converted.

- i_a, i_b and i_c from the transform is fed to the inverter with unity gain.

- Signal from the inverter is fed to machine side which has three phase to two phase transform block.

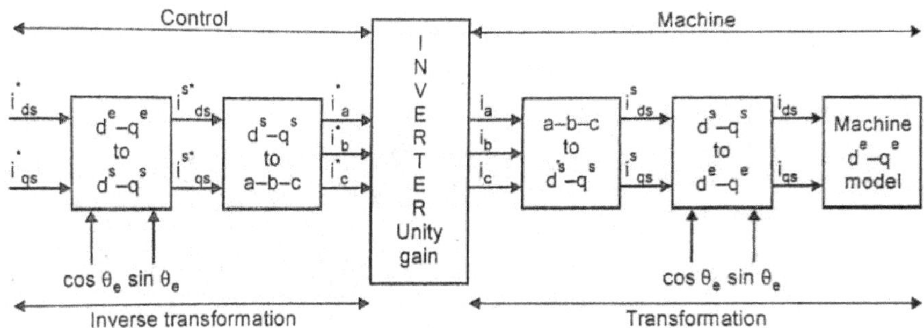

Fig. 3.35. Vector Control Implementation in Induction Motor

- Then the direct axis and quadrature axis stator current is converted into model and fed to the machine model.

- Cos θ_e and Sin θ_e are the reference signals during the above conversion.

Features of Vector Control

- High transient current capability

- Increased speed range

- Lower inertia

3.9. APPLICATIONS

SQUIRREL CAGE INDUCTION MOTOR

- ✓ Lathes and drilling machines
- ✓ Fans and blowers
- ✓ Water pump and grinders etc.,

SLIP RING INDUCTION MOTOR

- ✓ Lifts
- ✓ Hoists
- ✓ Cranes and elevators
- ✓ Compressors

STATOR VOLTAGE CONTROL

- ✓ Low power applications like fans, blowers and centrifugal pumps
- ✓ Used for starting high power induction motors.

V/F CONTROL

- ✓ Traction
- ✓ Steel mills
- ✓ Pumps and fans
- ✓ Blowers, conveyors and compressors
- ✓ Machine tool, etc.,

CYCLOCONVERTER CONTROL

- ✓ Ball mill in a cement plant

SLIP POWER RECOVERY CONTROL

- ✓ Variable speed wind energy systems
- ✓ Large capacity wind energy systems
- ✓ Variable speed hydro pump/generator

PROBLEMS

1. **A 3-phase slip ring Induction motor has chopper controlled resistance in the rotor circuit for speed control with the chopper completely ON always, the maximum torque occurs at a slip of 0.2. With chopper completely OFF the maximum torque occurs at a slip of 1. Determine the value of resistance in the chopper.**

Solution:

Neglecting Stator Impedance,

Slip when chopper is

Completely ON, $\qquad S_1 = 0.2$

Slip when chopper is

Completely OFF, $\qquad S_2 = 1$

$$\frac{S_2}{S_1} = \frac{R_2 + R_e}{R_2} = \frac{1}{0.2} = 5.$$

$\Rightarrow$ $R_e = 4R_2$.

If R is the resistance in the chopper, the copper loss is,

$$= 3I_2^2 R_2 + \left[R_d + (1-\alpha)R \right] I_d^2$$

But, $\quad \dfrac{I_2}{I_d} = \sqrt{\dfrac{2}{3}}$

$$\text{Copper loss, } = 3I_2^2 \left[R_2 + \frac{1}{2}(1-\alpha)R \right]$$

$$= 3I_2^2 \left[R_2 + R_e \right]$$

$$\Rightarrow R_e = \frac{1}{2}[1-\alpha]R.$$

(i) When chopper is completely ON,

$$\frac{T_{ON}}{T} = 1, \Rightarrow R_e = \frac{1}{2}[1-1]R = 0$$

(ii) When chopper is completely OFF,

$$\frac{T_{ON}}{T} = 0, \quad => R_e = \frac{1}{2}[1-0]R = 0.5R$$

Substituting,

$$R_e = 4\,R_2$$

$$=> \quad 0.5R = 4\,R_2$$

$$=> \quad R = 8\,R_2.$$

A resistance $R = 8R_2$ must be placed so that the Torque-speed curve will have its slip for maximum torque at unity, if the chopper is always OFF.

2. A 3-phase, 4-pole, 50 Hz Induction motor has a chopper controlled resistance in the rotor circuit for speed control. Load torque is ω^2, when the thyristor is ON, the torque is 30 N-m at a slip of average 0.03. If T_{ON} / T_{OFF} = 1, compute the average torque and speed. The motor develops a torque of 80 percent of ON torque when the thyristor is OFF. The speed variation ranges down to 1200 rpm from synchronous speed. Determine the ratio of T_{ON}/ T_{OFF} to give an average torque of 25 N-m.

Solution:

When thyristor is continuously ON torque is = T = 30 N-m.

When thyristor is Continuously OFF torque is

80 % of ON torque = T = 24 N-m.

The synchronous speed, N_S = 1500 rpm.

Slip,   S = 0.03.

The speed of the motor at this slip, N = (1-0.03) 1500

$$= 1465 \text{ rpm.}$$

When thyristor chopper is OFF, the complete resistance is included in the circuit.

When T_{ON} / T_{OFF} = 1,

Average torque, $\text{T}_{ar} = \dfrac{30\times1+24\times1}{2} = 27$

Since $\text{T}\,\alpha\,\text{N}^2$, therefore

$$\frac{T_1}{T_2} = \frac{N_1^2}{N_2^2}$$

(or) $\qquad \dfrac{T_1}{N_1^2} = \dfrac{T_2}{N_2^2}$

$$\frac{30}{(1465)^2} = \frac{27}{N_2^2}$$

Speed when T_{ON} / T_{OFF} = 1,

$$\text{N}_2 \quad = \frac{27}{30}\times(1465)^2$$
$$= 1390 \text{ rpm.}$$

When speed is 1200 rpm for chopper OFF, torque would be

$$\frac{(1200)^2}{(1465)^2}\times30 = 20.13\,N-m.$$

Average torque = 25 N-m.

$$25 \quad = \frac{30\times T_{ON} + 20.13 T_{OFF}}{T_{ON} + T_{OFF}}$$

$$25 \quad = \frac{30\times\dfrac{T_{ON}}{T_{OFF}} + 20.13}{\dfrac{T_{ON}}{T_{OFF}} + 1}$$

$$\Rightarrow \quad \left[\frac{T_{ON}}{T_{OFF}} + 1\right]25 = 30\times\frac{T_{ON}}{T_{OFF}} + 20.13$$

$$\Rightarrow \quad \frac{T_{ON}}{T_{OFF}} = 0.97$$

UNIT 4
SYNCHRONOUS MOTOR DRIVE

4.1. VARIABLE FREQUENCY CONTROL (OR) VARIABLE SPEED OPERATION

✓ In synchronous machine, the speed is directly proportional to the frequency. So the motor speed can be controlled by varying the frequency.

✓ Modes of variable frequency control are,

A) TRUE SYNCHRONOUS MODE (or) SEPARATE CONTROL MODE (or) OPEN LOOP V/F CONTOL

- Here, the stator supply frequency is controlled from initial to desired value, so that the difference between synchronous speed and rotor speed is always small.

- Variable frequency control method does, speed control, starting and regenerative braking.

- This type of control is used in control of multiple synchronous reluctance or permanent magnet motors which is used in fiber spinning, textile and paper mills.

- The system consists of rectifier, LC filter, PWM inverter, synchronous machines which is to be controlled, flux control block and delay circuit.

- AC Supply is converted into DC supply by the rectifier and the ripples produced during the conversion is eliminated by the LC filter and fed to the inverter.

- Inverter converts the DC supply into AC supply and feds to the parallel connected synchronous machines.

- Here the machine has damper winding to prevent the oscillations.

- Now to control the speed of the machine, command frequency is applied to the delay circuit block, inverter and finally to the rectifier by means of flux control block.

- Flux control block is used to maintain constant flux for below the base speed and constant terminal voltage for above the base speed.

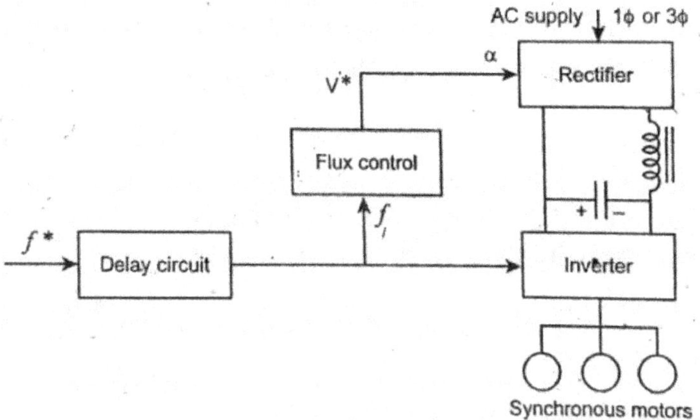

Fig.4.1. Open Loop V/F Control Mode

B) SELF CONTROL MODE

- Here the stator winding of the machine is connected with the inverter, where this inverter supplies variable frequency variable voltage sinusoidal supply.

- The frequency, phase position of stator volatge and torque angle is sensed by rotor position encoder and fed to the control unit.

- The control unit also receives the delay command.

- The control block generates firing pulses for the inverter according to the signal received from the rotor position encoder and the delay command.

- Here, the firing signals fed to inverter used for both control and commutation of the inverter, so it is known as load commutated inverter (LCI).

- If in case, the synchronous motors are over excited, they can supply reactive power which can be used for commutation of the thyristors in the inverter, so it called as line commutated inverter.

- The self-controlled motor is similar to that of DC motor under both steady state and dynamic condition, so it is called as commutator less motor (CLM).

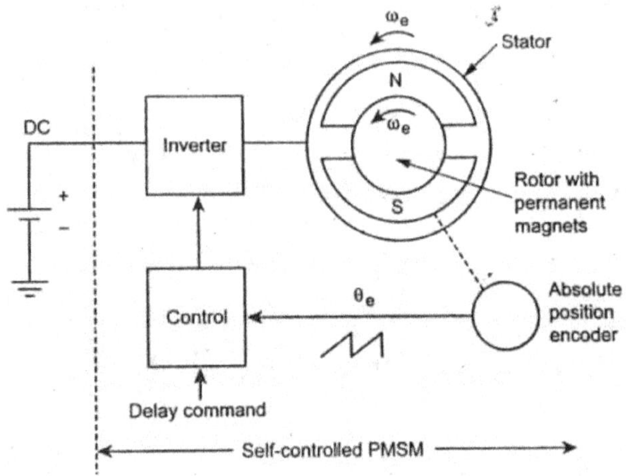

Fig.4.2. Self-Control Mode

4.2. CONSTATNT MARGINAL ANGLE CONTROL

- Constant marginal angle control is one of the best method to use inverter and the motor in efficient manner.

- Maximum torque and highest power factor operation can be achieved by operating the inverter at minimum permissible value of margin angle.

- The drive has outer speed loop and inner current loop.

- Rotor position encoder senses the actual rotor position (ω_m) and it is compared with the reference speed (ω_m^*) in the comparator.

- Thus the comparator generates error signal and it is fed to the speed controller and the current limiter.

- This current limiter gives out the reference current (I_d^*).

- DC link current (I_d) is taken and fed to the comparator, it compares reference current (I_d^*) and DC link current (I_d).

- The signal developed by the comparator is fed to the current controller and firing circuit.

- It generates the firing pulses accordingly and feds to the controlled rectifier circuit.

- From the value of DC link current command I_d^*, I_s and 0.5u are produced by block (1) & (2).

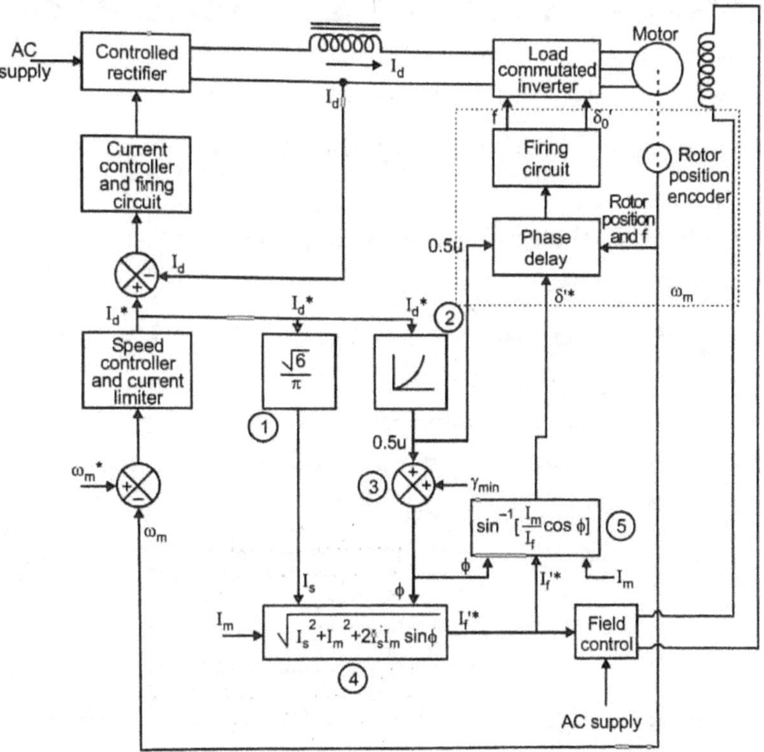

Fig.4.3. Constant margin angle control

- The Φ signal in a adder (3) is generated by adding 0.5u and γ_{min}.

- In the block (4), I_f^{l*} is calculated from I_s, 0.5u and I_m.

- I_m (Magnetizing current) is held constant to keep the flux constant.

- Block (5) calculates δ^{l*} from the known values Φ and I_f^{l*}

- Phase delay circuit shifts the pulses produced by the encoder to generate desired value of δ_0^l and F.

- It is fed to load commutated inverter for turning on and off of the thyristor to control the machine.

4.3. POWER FACTOR CONTROL

- The main aim of this method of control is controlling the power factor by varying the field current.

- This type of control is done in wound field machines.

- If the motor is operated at unity power factor, the current drawn by the motor would have lowest magnitude therefore, there occurs lowest internal copper loss.

- Here, in this control, voltage and current are sensed and fed to the power factor calculator.

- The power factor calculator calculates the phase angle between the current and the voltage, which is nothing but the power factor (P=V*I).

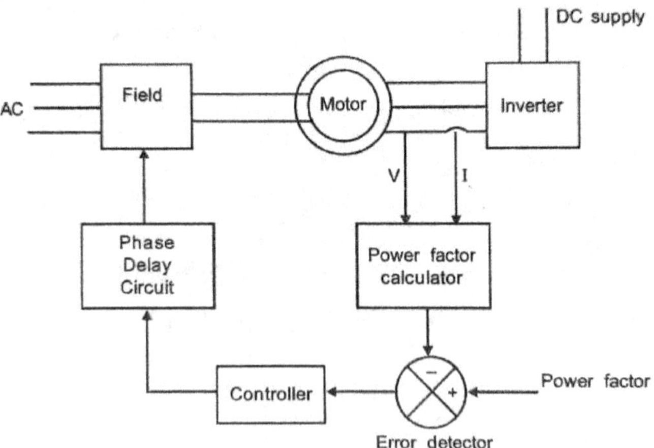

Fig.4.4. Power Factor Control

- The calculated power factor value is compared with the reference power factor value in comparator, thus generates the error signal.

- The error signal is fed to the controller and then to the phase delay circuits and then finally to the field block.

- The signal fed is to generate the appropriate power factor signal as per the commanded value.

4.4. THREE PHASE VOLTGE SOURCE INVERTER (VSI) FED SYNCHRONOUS MOTOR DRIVES

- VSI is also the type of self-control drive, which uses the rotor position sensor.

- In this mode crystal oscillator is used to determine the speed of the motor.

- In the 180^0 conduction mode of VSI, load commutation cannot be achieved.

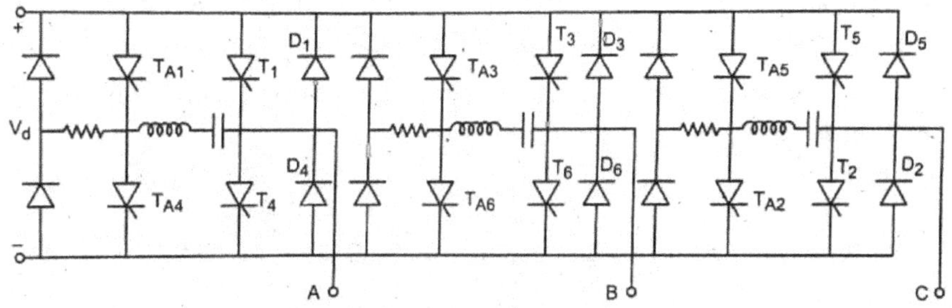

Fig.4.5. Voltage Source Invereter

- So the forced commutation is done.

- Variable voltage variable frequency input to the motor can be obtained by the following methods,

a) **Square wave inverter** – Consists of phase controlled rectifier, filter and the square wave inverter, as its output waveform is of square shape. Here the speed is sensed and corresponding frequency and voltage signal is developed and fed to the rectifier and the inverter respectively to control the switches and then the motor.

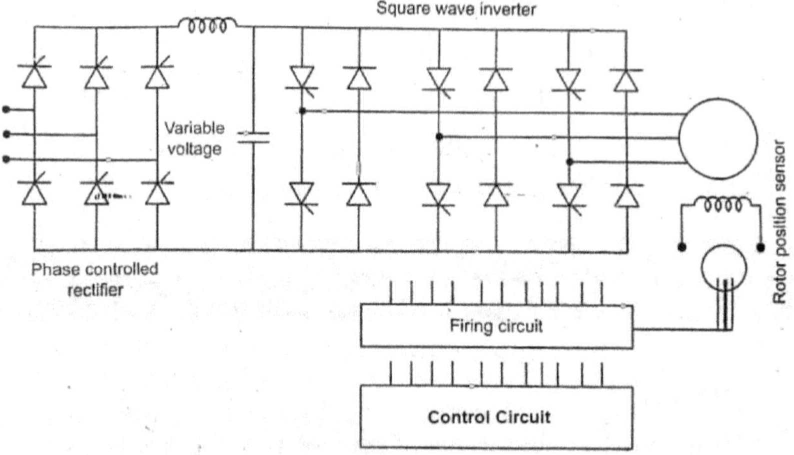

Fig.4.6. Self-controlled Square wave inverter

b) **PWM Inverter** – Consists of uncontrolled rectifier, filter and the PWM inverter. PWM technique provides voltage control. Here the DC link voltage is maintained constant by using the diode rectifier. Here the speed is sensed and the correspondingly voltage and frequency signal is developed and fed to the inverter.

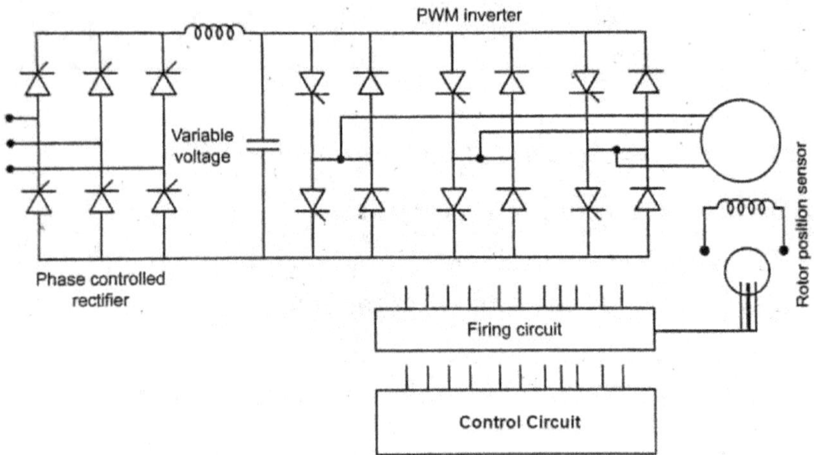

Fig.4.7. Self-controlled PWM inverter

c) **Chopper with square wave inverter** – Consists of diode rectifier, chopper, filter and the square wave inverter. Chopper is fixed in between the diode rectifier and the filter. Here some simple converter is used to reduce the link inductance.

- If the inverter used in the system is square wave inverter, stator current has sharp peaks, which has high harmonics produces pulsating torque, causes reduced performance and additional heat.

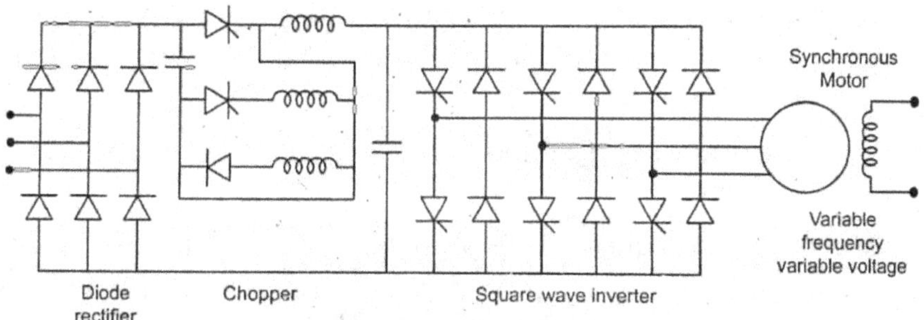

Fig.4.8. Chopper fed square wave inverter

- But if the inverter is PWM inverter, stator current has low peaks, which has low harmonics, torque pulsations.

- Power factor in square wave inverter lags but in case of PWM inverter, power factor is unity.

- In order to get inverter with low losses and reduced size, inverter with unity power factor must be choosed.

Advantages:

1. VSI drive provides good efficiency
2. Multi-motor operation is possible
3. PWM drive has better dynamic response

Disadvantages:

1. The converter cost is high

2. By using VSI there is instability and harmonic content in the output

3. VSI is not applicable for low speed operations

4.5. THREE PHASE CURRENT SOURCE INVERTER (CSI) FED SYNCHRONOUS MOTOR DRIVES

- CSI can have either self-control or separate control, but due to the stability, self-control is employed.

- When the synchronous motor is synchronous fed, it operates in leading power factor, so the machine voltage is used for the commutation and it is known as load commutation.

- Load commutated CSI fed synchronous motor is known as converter motor.

- Machine starts above 10% of its base speed due to machine commutation, but by forced commutation, machine starts from 0% of the base speed.

- Cost of inverter increases due to forced commutation.

- CSI system has moderate efficiency.

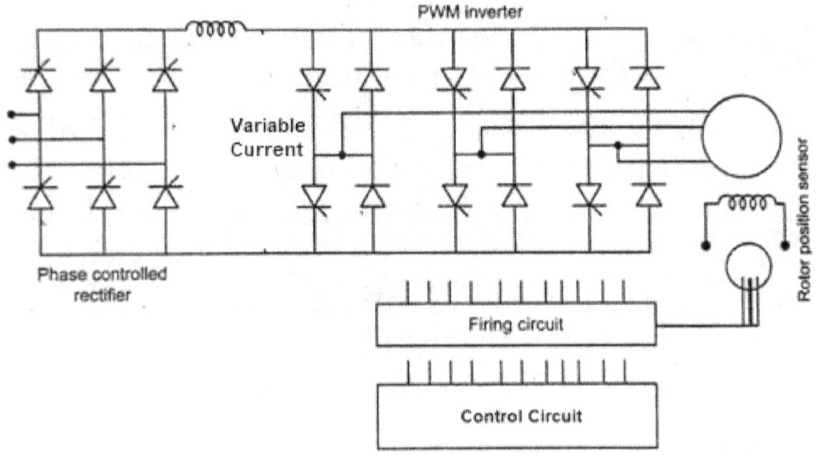

Fig.4.9. Self-controlled PWM inverter

- There may be voltage spikes during the commutation which affects the insulation.

- Damper winding is used in the machine to limit the voltage spikes.

- If the inverter used in the system is square wave inverter, stator current has sharp peaks, which has high harmonics produces pulsating torque, causes reduced performance and additional heat.

- But if the inverter is PWM inverter, stator current has low peaks, which has low harmonics, torque pulsations.

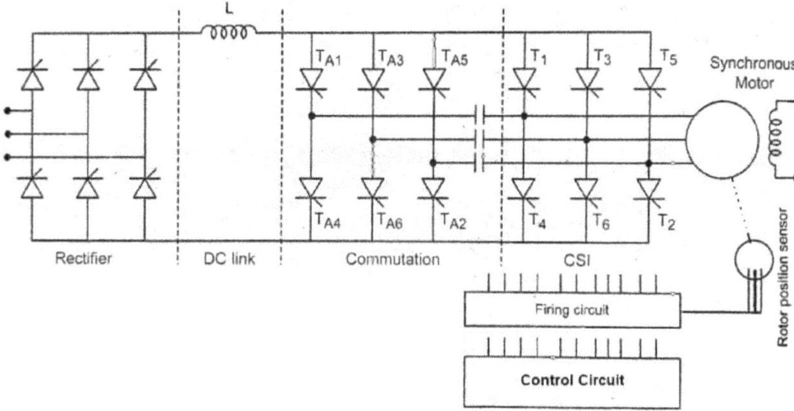

Fig.4.10. Self-controlled square wave inverter with forced commutation

Advantages:

1. CSI is robust and simple

2. Four quadrant operation is possible

3. Speed control is simple

4. Better controlled performance.

Disadvantages:

1. Cost is expensive with PWM technique

2. Unsupported for multi-motor operation

3. CSI not used in open loop control drives

4. Undesirable dynamic performance

5. At no-load condition it is difficult to operate.

4.6. CYCLOCONVERTER FED SYNCHRONOUS MACHINE DRIVE

- VSI and CSI fed synchronous motor drive have 2 stage of conversion to produce variable frequency and variable voltage.

- But cycloconverter fed synchronous motor drive has only one stage of conversion to produce the same.

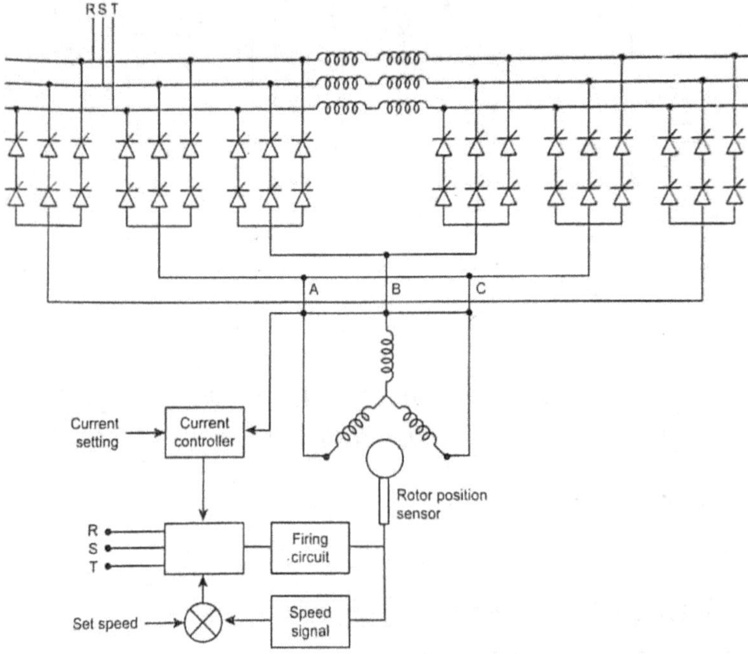

Fig.4.11. Cycloconverter fed synchronous motor drive

- Line voltage is used to commutate the thyristors of cycloconverter.

- The output frequency can be varied from 0 to 1/3 of the input frequency.

- Speed control is limited from 0 to 1/3 of its base speed.

- Power transfer is of both the direction in cycloconverter fed drive.

- In high speed ranges, cycloconverter gives high quality sinusoidal output voltage.

- Power factor in this drive is unity.

- This drive provides good efficiency and good dynamic behavior.

- Cycloconverter circuit is complex as it have lager number of thyristors. Cost is high due to large number of thyristors.

- Cycloconverter fed drive is used in reversing mines, hoists, etc.,

4.7. APPLICATIONS

1) SYNCHRONOUS MOTOR DRIVES

- Ball mills, Robot actuators, Textile mills, Saw mills, Spinning mills, Conveyor belts, etc.,

2) VARIABLE FREQUENCY DRIVES
a) *Separate control*

- Paper mills, Saw mills, Textile mills, Spinning mills, etc.,

b) *Self-control*

- A self-controlled synchronous motor is a similar to that of D.C. motor drive and it is used where a D.C. motor is objectionable due to its mechanical Commutator, which limits the speed range and power output.

3) CONSTANT MARGIN ANGLE CONTROL

- High power drives

- High speed drives like blowers, compressors, steel mills, induced and forced draft fans, flywheel energy storage, etc.,

4) CSI & VSI FED SYNCHRONOUS MOTOR DRIVES

- CSI is used in drives such as, gas turbine starting, pumped hydro turbine starting, pump, blower drives, etc.,

- VSI used for high and medium power applications, PWM used for low and medium power applications.

4.8. CLOSED LOOP CONTROL OF SYNCHRONOUS MACHINE DRIVE

- This system consists of outer speed control loop and inner current control loop.

- The terminal voltage sensor senses the terminal voltage of the machine and generates frequency and phase signal, feds to the phase delay block.

- The fed signal acts as a reference pulses. Phase delay circuit shifts the reference pulses suitably to control the converter with constant commutation lead angle (β_{ic}).

- The actual speed signal is compared with the reference speed to generate the error signal.

- Depending on the sign of speed error signal, the β_{ic} is set to provide motoring or braking mode of operation.

- An increase in reference speed produces positive speed error. So that β_{ic} is set to provide motoring operation.

- A decrease in reference speed produces negative speed error. So that β_{ic} is set to provide braking operation.

- The speed error signal generated by the comparator is fed to the speed controller and then to the current limiter to generate reference DC link current.

- Now the reference DC link current and the actual DC link current is compared in the comparator which generates control signal through current controller and firing circuit to control the source side converter.

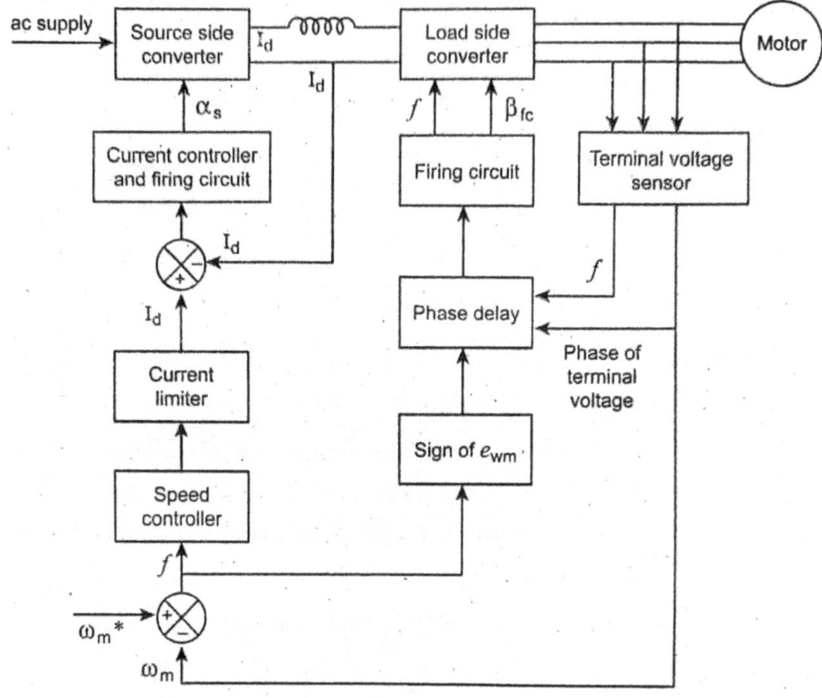

Fig.4.12. Closed loop control

Advantages

- High efficiency
- Ability to run at run speeds
- Four quadrant operations with regenerative braking, etc.,

Applications

- High speed and high power drives for compressors, blowers, fans, pumps, conveyors, steel rolling mills, ship propulsion etc.,

UNIT 5
DESIGN OF CONTROLLERS FOR DRIVES

5.1. TRANSFER FUNCTION OF DC MOTOR AND LOAD

- The speed of DC motor is directly proportional to armature voltage and inversely proportional to flux in field winding.
- Here the desired speed is obtained by varying the armature voltage.
- This speed control system is an electro-mechanical control system.
- Here, the transfer function of armature controlled dc motor and load is to be discussed.
- The electrical system consists of the armature and the field circuit but for analysis, only the armature circuit is considered because the field is excited by a constant voltage.
- The mechanical system consist of the rotating part of the motor and load connected to the shaft of the motor.
- The armature controlled DC motor speed control system is shown in the below figure.

Let,

R_a – Armature resistance

L_a – Armature inductance

I_a – Armature current

V_a – Armature voltage

e_b – Back emf

K_t – Torque constant

T – Torque developed by motor

θ - Angular displacement of the shaft

J – Moment of inertia

B – Friction co-efficient of motor and load

K_b – Back emf constant

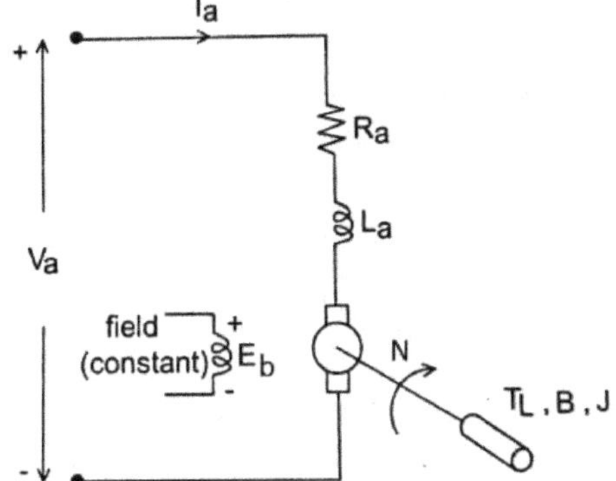

Fig.5.1. Equivalent circuit of separately excited DC motor

Kirchoffs voltage law is given as,

$$V_a = R_a i_a + L_a \frac{di_a}{dt} + e_b \text{ ----(1)}$$

Where, $e_b \, \alpha \, \dfrac{d\varnothing}{dt}$

$$e_b = K_a \Phi n \text{ ----(2)}$$

Torque equation,

$$T = T_L + T_B + T_J$$

$$T = T_L + B\frac{d\theta}{dt} + J\frac{d^2\theta}{dt}$$

$$T = T_L + Bn + J\frac{dn}{dt} \text{ ----(3)}$$

WKT, $T = K_a \Phi i_a$ ----(4)

Apply laplace transform in (1), (2), (3) & (4)

(1) becomes, $V_a(S) = R_a I_a(S) + SL_a I_a(S) + E_b(S)$ ----(5)

(2) becomes, $E_b(S) = K_a \Phi N(S)$ ----(6)

(3) becomes, $T(S) = T_L(S) + BN(S) + JSN(S)$ ----(7)

(4) becomes, $T(S) = K_a \Phi I_a(S)$ ----(8)

(5) => $V_a(S) - R_a I_a(S) + SL_a I_a(S) + E_b(S)$

$$V_a(S) = E_b(S) + R_a I_a(S) + SL_a I_a(S)$$

$$V_a(S) = E_b(S) + I_a(S)[R_a + SL_a]$$

$$V_a(S) - E_b(S) = I_a(S)[R_a + SL_a]$$

$$I_a(S) = \frac{V_a(S) - E_b(S)}{R_a + SL_a} \Rightarrow \frac{V_a(S) - E_b(S)}{R_a\left[1 + S\frac{L_a}{R_a}\right]}$$

$$I_a(S) = \frac{V_a(S) - E_b(S)}{R_a[1 + S\tau_a]} \quad ----(9)$$

Where, $\tau_a = \dfrac{L_a}{R_a}$

$$(7) \Rightarrow T(S) = T_L(S) + BN(S) + JSN(S)$$

$$T(S) = T_L(S) + N(S)[B + JS]$$

$$T(S) - T_L(S) = N(S)[B + JS]$$

$$N(S) = \frac{T(S) - T_L(S)}{B + JS}$$

$$N(S) = \frac{T(S) - T_L(S)}{B\left[1 + \frac{J}{B}S\right]}$$

$$N(S) = \frac{T(S) - T_L(S)}{B[1 + S\tau_m]} \quad ----(10)$$

Where, $\tau_m = \dfrac{J}{B}$

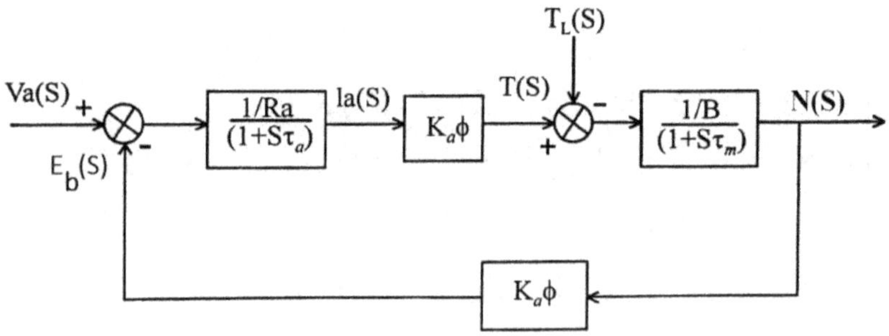

Fig.5.2. Transfer function of separately excited DC motor

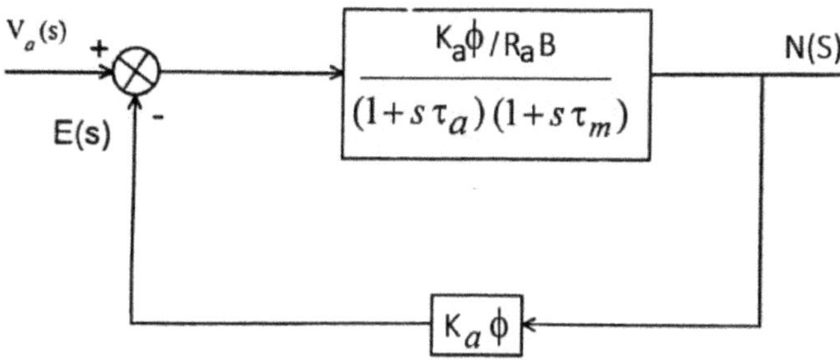

Fig.5.3. Transfer function of separately excited DC motor with $T_L(S)$ neglection

$$\frac{N(S)}{V_a(S)} = \frac{\dfrac{K_a\Phi}{R_aB}}{(1+S\tau_m)(1+S\tau_a)} \Bigg/ \left(1 + \frac{\dfrac{K_a\Phi}{R_aB}}{(1+S\tau_m)(1+S\tau_a)} * K_a\Phi\right)$$

$$\frac{N(S)}{V_a(S)} = \frac{\dfrac{K_a\Phi}{R_aB}}{\cancel{(1+S\tau_m)(1+S\tau_a)}} \Bigg/ \frac{(1+S\tau_m)(1+S\tau_a)+\dfrac{K_a\Phi}{R_aB}}{\cancel{(1+S\tau_m)(1+S\tau_a)}} * K_a\Phi$$

$$\frac{N(S)}{V_a(S)} = \frac{\dfrac{K_a\Phi}{\cancel{R_aB}}}{\dfrac{(1+S\tau_m)(1+S\tau_a)R_aB+(K_a\Phi)^2}{\cancel{R_aB}}}$$

$$\frac{N(S)}{V_a(S)} = \frac{K_a\Phi}{(1+S\tau_m)(1+S\tau_a)R_aB+(K_a\Phi)^2}$$

If $\tau_a \ll \tau_m$, τ_a is neglected

So, $$\frac{N(S)}{V_a(S)} = \frac{K_a\Phi}{(1+S\tau_m)R_aB+(K_a\Phi)^2}$$

$$\frac{N(S)}{V_a(S)} = \frac{K_a\Phi}{(K_a\Phi)^2 + R_aB + S\tau_m R_aB}$$

$$\frac{N(S)}{V_a(S)} = \frac{K_a\Phi}{(K_a\Phi)^2 + R_aB\left[1+\dfrac{S\tau_m R_aB}{(K_a\Phi)^2 + R_aB}\right]}$$

$$\frac{N(S)}{V_a(S)} = \frac{K_m}{1+S\tau_{m1}}$$

Where, $K_m = \dfrac{K_a}{R_aB+(K_a\Phi)^2}$

$$\tau_{m1} = \frac{R_aB\tau_m}{R_aB+(K_a\Phi)^2}$$

$$\frac{N(S)}{I_a(S)} = \frac{K_{m2}}{1+S\tau_m}$$

$$\frac{I_a(S)}{V_a(S)} = \frac{N(S)}{V_a(S)} * \frac{I_a(S)}{N(S)}$$

$$= \frac{K_m}{1+S\tau_{m1}} * \frac{1+S\tau_m}{K_{m2}}$$

$$= \frac{K_m}{1+S\tau_{m1}} * \frac{1+S\tau_m}{\dfrac{K_a\Phi}{B}}$$

Where, $K_{m2} = \dfrac{K_a\Phi}{B}$

$$= \frac{K_m(1+S\tau_m)}{\dfrac{K_a\Phi}{B}(1+S\tau_{m1})}$$

$$= \frac{K_mB(1+S\tau_m)}{K_a\Phi(1+S\tau_{m1})}$$

$$\frac{I_a(S)}{V_a(S)} = \frac{K_{m1}(1+S\tau_m)}{(1+S\tau_{m1})}$$

Where, $K_{m1} = \dfrac{K_mB}{K_a\Phi}$

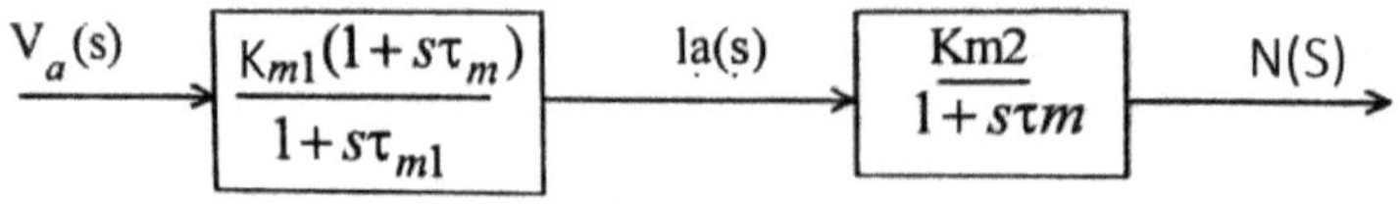

$$V_a(s) \longrightarrow \boxed{\dfrac{K_{m1}(1+s\tau_m)}{1+s\tau_{m1}}} \xrightarrow{\ Ia(s)\ } \boxed{\dfrac{Km2}{1+s\tau m}} \longrightarrow N(S)$$

Fig.5.4. Block diagram of motor

5.2. TRANSFER FUNCTION OF CONVERTER

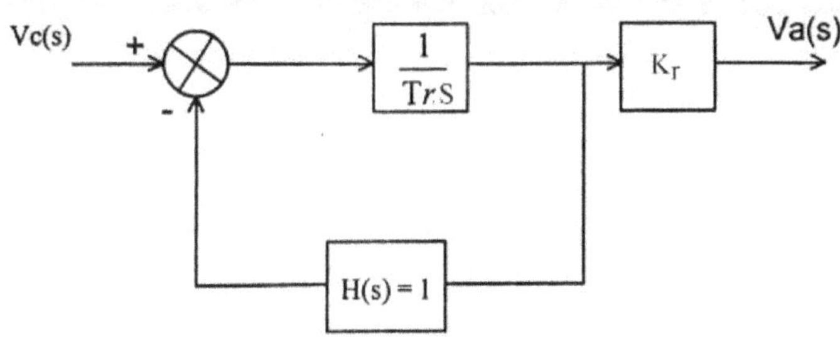

Fig.5.5. Transfer function of power converter

$$\frac{V_a(S)}{V_C(S)} = \frac{\dfrac{1}{T_rS}}{\left(1+\dfrac{1}{T_rS}\right)*1} * K_r$$

$$\frac{V_a(S)}{V_C(S)} = \frac{\dfrac{1}{T_rS}}{\dfrac{T_rS+1}{T_rS}} * K_r$$

$$\frac{V_a(S)}{V_C(S)} = \frac{K_r}{1+ST_r}$$

TF of current controller, $G_C(S) = \dfrac{K_C(1+ST_C)}{ST_C}$

TF of speed controller, $G_S(S) = \dfrac{K_S(1+ST_S)}{ST_S}$

Where,

Kc & Ks – Current controller & speed controller gain.

Tc & Ts – Time constant of the controller

Gain of the converter, $Kr = 1.35\dfrac{V_C}{V_{cm}}$

Where,

$\dfrac{V_C}{V_{cm}}$ – Normalized control voltage

5.3. CLOSED LOOP CONTROL WITH CURRENT AND SPEED FEEDBACK

- This drive consists of inner current loop and outer speed loop.

- The inner current loop is used to make motor current/ torque within the safer value and it also improves the drive operation during nonlinear operation of the converter.

- Speed of the machine (ω_{m_act}) is sensed by the speed sensor and compared with the reference value (ω_{m_ref}) to generate the error speed signal (ωerr) and fed to the speed controller block.

- Speed controller generates the control signal ω_c and feds to the current limiter.

- Current limiter generates reference current signal (I_{ref}).

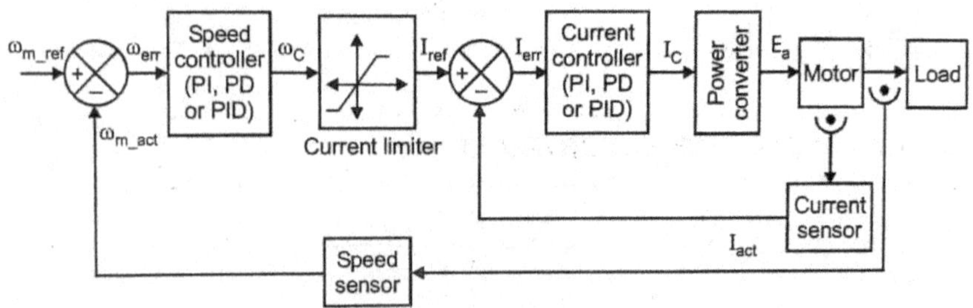

Fig.5.6. Closed loop with current and speed feedback

- If in case, the speed error signal is positive, reference current signal (I_{ref}) increases. Current limiter saturates. This leads to acceleration of motor.

- So, the motor speed is made within the permissible value.

- If in case, the speed error signal is negative, reference current signal (I_{ref}) decreases. Current limiter saturates. This leads to deceleration of motor.

- So, the motor speed is made within the desired value.

- Now the reference current (I_{ref}) from the current limiter is compared with the motor current sensed by current sensor in the comparator to generate current error signal (I_{err}).

- The current error signal is fed to the current controller to generate control signal (Ic).

- This control signal (Ic) controls the semiconductor switches in the power converter to control the speed of the motor.

5.4. ARMATURE VOLTAGE CONTROL AND FIELD WEAKENING CONTROL

- This drive employs inner current control loop and outer speed loop.

- This drive operates,

    - At constant field current and variable armature voltage below base speed.

    - At constant armature voltage and variable field current above the base speed.

Operation below base speed

✓ Here, in field control loop, the back emf E is compared with the reference voltage E* (0.85 to 0.95 of the rated voltage).

✓ Comparator generates error signal, in which it is fed into the field controller and then to firing controller.

✓ If the error is of larger value, controller saturates.

✓ Firing controller maintains the firing pulse as 0, if the rated voltage is applied to the motor.

✓ At the same time, speed ω_m is sensed by the speed sensor and it is compared with the reference speed ω_m^*.

✓ The error signal is fed to the speed controller and then to the current limiter.

✓ Current limiter generates the reference current signal I_a^*.

✓ If in case speed error signal is positive, reference current signal I_a^* increases. Current limiter saturates. This leads to acceleration of motor.

✓ So, the current signal I_a^* is limited and the motor speed is made equal to the reference speed.

✓ If in case speed error signal is negative, deceleration of motor occurs, so I_a^* is limited.

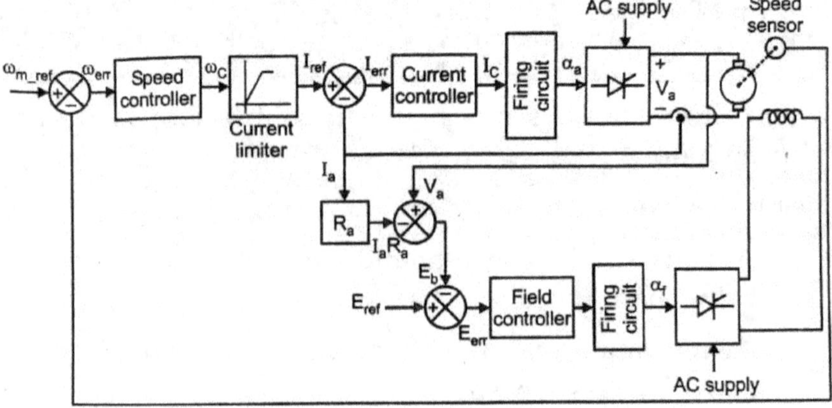

Fig.5.7. Armature Voltage Control & Field Weakening Control with closed loop speed control of DC Drive

Operation above base speed

✓ When the motor running at speed closer to the base speed, armature voltage V_a is nearer to the rated value and field controller comes out of the saturation.

✓ If there is occurrence of speed error signal, Va increases, motor accelerates, E increases, field error e_f decreases, reducing the field current.

✓ So the motor decelerates till it reaches the reference speed.

5.5. DESIGN OF CONTROLLERS

(i) SPEED CONTROLLER

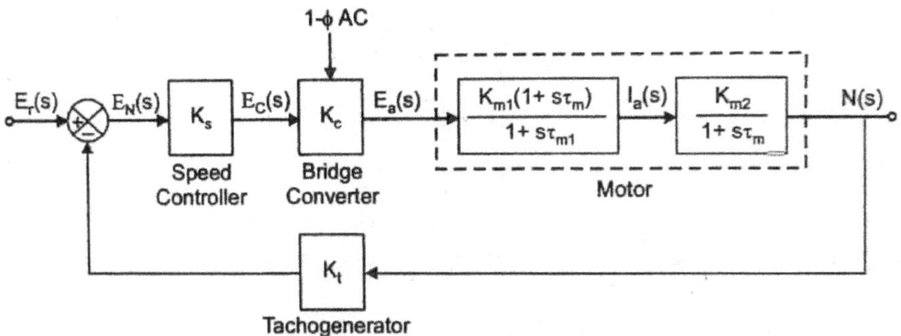

Fig.5.8. Speed Control Loop

$$\frac{N(S)}{E_r(S)} = \frac{G(S)}{1+G(S)H(S)} \quad ----(1)$$

$$G(S) = Ks.Kc.\frac{K_{m1}(1+S\tau_{m1})}{1+S\tau_{m1}} . \frac{K_{m2}}{1+S\tau_m}$$

$$G(S) = \frac{K_S K_C K_{m1} K_{m2}\cancel{(1+S\tau_m)}}{(1+S\tau_{m1})\cancel{(1+S\tau_m)}}$$

$$G(S) = \frac{K_S K_C K_{m1} K_{m2}}{1+S\tau_{m1}} \quad ----(2)$$

$$H(S) = K_t \quad ----(3)$$

Subs (2) & (3) in (1)

$$\frac{N(S)}{E_r(S)} = \frac{\dfrac{K_S K_C K_{m1} K_{m2}}{1+S\tau_{m1}}}{1+\dfrac{K_S K_C K_{m1} K_{m2} K_t}{1+S\tau_{m1}}}$$

$$\frac{N(S)}{E_r(S)} = \frac{\dfrac{K_S K_C K_{m1} K_{m2}}{\cancel{1+S\tau_{m1}}}}{\dfrac{1+S\tau_{m1}+K_S K_C K_{m1} K_{m2} K_t}{\cancel{1+S\tau_{m1}}}}$$

$$\frac{N(S)}{E_r(S)} = \frac{K_S K_C K_{m1} K_{m2}}{1+S\tau_{m1}+K_S K_C K_{m1} K_{m2} K_t} \quad ----(4)$$

Div. $(1+K_S K_C K_{m1} K_{m2} K_t)$ in num. and den. of (4)

$$\frac{N(S)}{E_r(S)} = \frac{\dfrac{K_S K_C K_{m1} K_{m2}}{1+K_S K_C K_{m1} K_{m2} K_t}}{1+\dfrac{S\tau_{m1}+K_S K_C K_{m1} K_{m2} K_t}{1+K_S K_C K_{m1} K_{m2} K_t}}$$

$$\frac{N(S)}{E_r(S)} = \frac{K_1}{1+S\tau_1} \quad\text{----(5)}$$

Where, $K_1 = \dfrac{K_S K_C K_{m1} K_{m2}}{1+K_S K_C K_{m1} K_{m2} K_t}$ ----(5.A)

$$\tau_1 = \frac{\tau_{m1}}{1+K_S K_C K_{m1} K_{m2} K_t} \quad\text{----(5.B)}$$

In (5.A) & (5.B), $K_S K_C K_{m1} K_{m2} K_t \gg 1$, so 1 is neglected.

(5.A) becomes, $K_1 = \dfrac{\cancel{K_S K_C K_{m1} K_{m2}}}{\cancel{K_S K_C K_{m1} K_{m2}} K_t} = \dfrac{1}{K_t}$ ----(5.C)

(5.B) becomes, $\tau_1 = \dfrac{\tau_{m1}}{K_S K_C K_{m1} K_{m2} K_t}$ ----(5.D)

WKT, $\dfrac{I_a(S)}{N(S)} = \dfrac{1+S\tau_{m1}}{K_{m2}}$ ----(6)

$$\frac{I_a(S)}{E_r(S)} = \frac{N(S)}{E_r(S)} * \frac{I_a(S)}{N(S)} \quad\text{----(7)}$$

Subs (6), (5) in (7)

$$\frac{I_a(S)}{E_r(S)} = \frac{K_1}{1+S\tau_1} * \frac{1+S\tau_m}{K_{m2}} => \frac{K_1(1+S\tau_m)}{K_{m2}(1+S\tau_1)}$$

$$I_a(S) = \frac{K_1(1+S\tau_m)}{K_{m2}(1+S\tau_1)} . E_r(S) \quad\text{----(8)}$$

If the input is step $E_r(S)$ in (8) becomes $\dfrac{E_r}{S}$

So, $I_a(S) = \dfrac{K_1(1+S\tau_m)}{K_{m2}(1+S\tau_1)} . \dfrac{E_r}{S}$

$$= \frac{K_1 E_r(1+S\tau_m)}{K_{m2} S(1+S\tau_1)} => \frac{K_1 E_r(1+S\tau_m)}{K_{m2} S\tau_1(\frac{1}{\tau_1}+S)} \quad\text{----(9)}$$

Div. $K_{m2}\tau_1$ in both num. & deno. of (9)

$$I_a(S) = \dfrac{\dfrac{K_1 E_r(1+S\tau_m)}{K_{m2}\tau_1}}{\dfrac{K_{m2}S\tau_1}{K_{m2}\tau_1}\left(\dfrac{1}{\tau_1}+S\right)} \Longrightarrow \dfrac{\dfrac{K_1 E_r(1+S\tau_m)}{K_{m2}\tau_1}}{S\left(\dfrac{1}{\tau_1}+S\right)} \quad ----(10)$$

Take partial fraction on (10)

$$\dfrac{\dfrac{K_1 E_r(1+S\tau_m)}{K_{m2}\tau_1}}{S\left(\dfrac{1}{\tau_1}+S\right)} = \dfrac{A_1}{S} + \dfrac{A_2}{S+\dfrac{1}{\tau_1}}$$

$$\dfrac{\dfrac{K_1 E_r(1+S\tau_m)}{K_{m2}\tau_1}}{S\left(\dfrac{1}{\tau_1}+S\right)} = \dfrac{A_1\left(S+\dfrac{1}{\tau_1}\right)+A_2 S}{S\left(S+\dfrac{1}{\tau_1}\right)}$$

$$\dfrac{K_1 E_r(1+S\tau_m)}{K_{m2}\tau_1} = A_1\left(S+\dfrac{1}{\tau_1}\right)+A_2 S$$

$$\dfrac{K_1 E_r+K_1 E_r S\tau_{m1}}{K_{m2}\tau_1} = A_1 S + \dfrac{A_1}{\tau_1} + A_2 S$$

$$\dfrac{K_1 E_r}{K_{m2}\tau_1} + \dfrac{K_1 E_r S\tau_{m1}}{K_{m2}\tau_1} = S(A_1+A_2) + \dfrac{A_1}{\tau_1} \quad ----(11)$$

Equate constants and S terms on both sides of (11)

$$\dfrac{K_1 E_r}{K_{m2}\tau_1} = \dfrac{A_1}{\tau_1} \Longrightarrow A_1 = \dfrac{K_1 E_r}{K_{m2}\tau_1}.\,\tau_1$$

$$A_1 = \dfrac{K_1 E_r}{K_{m2}} \quad ----(12)$$

$$\dfrac{K_1 E_r \tau_m}{K_{m2}\tau_1} = A_1+A_2 \quad ----(13)$$

Subs (12) in (13)

$$\dfrac{K_1 E_r \tau_m}{K_{m2}\tau_1} = \dfrac{K_1 E_r}{K_{m2}} + A_2$$

$$A_2 = \dfrac{K_1 E_r \tau_m}{K_{m2}\tau_1} - \dfrac{K_1 E_r}{K_{m2}}$$

$$A_2 = \dfrac{K_1 E_r}{K_{m2}}\left[\dfrac{\tau_m}{\tau_1} - 1\right] \quad ----(14)$$

$$I_a(S) = \frac{K_1 E_r}{K_{m2} S} + \left[\frac{K_1 E_r}{K_{m2}} * \frac{\frac{\tau_m}{\tau_1} - 1}{S + \frac{1}{\tau_1}}\right]$$

$$I_a(S) = \frac{K_1 E_r}{K_{m2} S} + \left[\frac{K_1 E_r}{K_{m2}} * \frac{\frac{\tau_m - \tau_1}{\tau_1}}{S + \frac{1}{\tau_1}}\right]$$

$$I_a(S) = \frac{K_1 E_r}{K_{m2} S} + \frac{K_1 E_r}{K_{m2} \tau_1} \frac{\tau_m - \tau_1}{S + \frac{1}{\tau_1}}$$

$$I_a(S) = \frac{K_1 E_r}{K_{m2}} \left[\frac{1}{S} + \frac{\tau_m - \tau_1}{\tau_1} * \frac{1}{S + \frac{1}{\tau_1}}\right] ----(15)$$

taking inverse laplace transform on above equ.

$$I_a(t) = \frac{K_1 E_r}{K_{m2}} \left[1 + \frac{\tau_m - \tau_1}{\tau_1} e^{\frac{-t}{\tau_1}}\right] ----(16)$$

$\tau_m >>>> \tau_1$, so τ_1 is neglected

So (16) becomes,

$$I_a(t) = \frac{K_1 E_r}{K_{m2}} \left[1 + \tau_m . e^{-t}\right] ----(17)$$

$$I_a(t) = I_a(\alpha); \ t = \alpha ----(18)$$

Subs (18) in (17)

$$I_a(\alpha) = \frac{K_1 E_r}{K_{m2}}$$

$$\frac{I_a(t)}{I_a(\alpha)} = \frac{\frac{K_1 E_r}{K_{m2}} \left[1 + \frac{\tau_m - \tau_1}{\tau_1} . e^{\frac{-t}{\tau_1}}\right]}{\frac{K_1 E_r}{K_{m2}}}$$

$$\frac{I_a(t)}{I_a(\alpha)} = 1 + \frac{\tau_m - \tau_1}{\tau_1} . e^{\frac{-t}{\tau_1}}$$

$\tau_m >>>> \tau_1$, so τ_1 is neglected

$$\frac{I_a(t)}{I_a(\alpha)} = 1 + \tau_m . e^{-t}$$

(ii) CURRENT CONTROLLER

a) P Controller

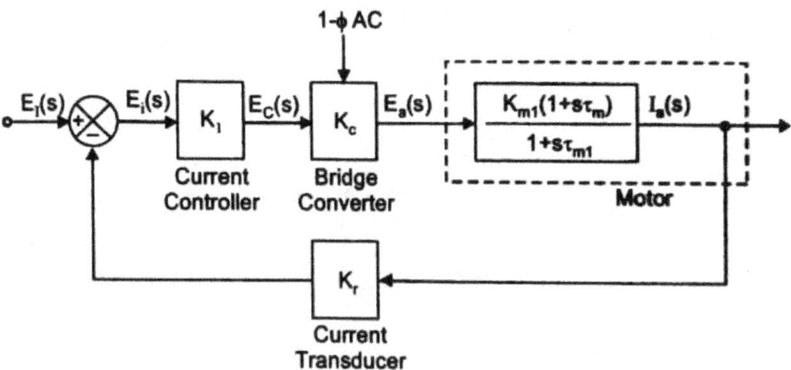

Fig.5.9. Current Control Loop

From the block diagram,

$$\frac{I_a(S)}{E_1(S)} = \frac{\dfrac{K_1 K_C K_{m1}(1+S\tau_m)}{1+S\tau_{m1}}}{1+\dfrac{K_1 K_C K_{m1}(1+S\tau_m)}{1+S\tau_{m1}}*K_r}$$

$$\frac{I_a(S)}{E_1(S)} = \frac{\dfrac{K_1 K_C K_{m1}(1+S\tau_m)}{1+S\tau_{m1}}}{\dfrac{(1+S\tau_{m1})+K_1 K_C K_{m1} K_r(1+S\tau_m)}{1+S\tau_{m1}}}$$

$$\frac{I_a(S)}{E_1(S)} = \frac{K_1 K_C K_{m1}(1+S\tau_m)}{(1+S\tau_{m1})+K_1 K_C K_{m1} K_r(1+S\tau_m)}$$

$$\frac{I_a(S)}{E_1(S)} = \frac{K_1 K_C K_{m1}(1+S\tau_m)}{1+S\tau_{m1}+K_1 K_C K_{m1} K_r+K_1 K_C K_{m1} K_r S\tau_m}$$

$$\frac{I_a(S)}{E_1(S)} = \frac{K_1 K_C K_{m1}(1+S\tau_m)}{1+K_1 K_C K_{m1} K_r+S\tau_{m1}+ K_1 K_C K_{m1} K_r S\tau_m}$$

$$\frac{I_a(S)}{E_1(S)} = \frac{K_1 K_C K_{m1}(1+S\tau_m)}{1+K_1 K_C K_{m1} K_r+S\tau_{m1}+ K_1 K_C K_{m1} K_r S\tau_m}$$

$$\frac{I_a(S)}{E_1(S)} = \frac{K_1 K_C K_{m1}(1+S\tau_m)}{1+K_1 K_C K_{m1} K_r+S(\tau_{m1}+ K_1 K_C K_{m1} K_r \tau_m)} \quad ----(1)$$

Divide $(1+K_1K_CK_{m1}K_r)$ in numerator & denominator of (1)

$$\frac{I_a(S)}{E_1(S)} = \frac{\dfrac{K_1K_CK_{m1}(1+S\tau_m)}{1+K_1K_CK_{m1}K_r}}{\left(\dfrac{\cancel{1+K_1K_CK_{m1}K_r}}{\cancel{1+K_1K_CK_{m1}K_r}}\right)+S\left(\dfrac{\tau_{m1}+K_1K_CK_{m1}K_r\tau_m}{1+K_1K_CK_{m1}K_r}\right)}$$

$$\frac{I_a(S)}{E_1(S)} = \frac{\dfrac{K_1K_CK_{m1}(1+S\tau_m)}{1+K_1K_CK_{m1}K_r}}{1+S\left(\dfrac{\tau_{m1}+K_1K_CK_{m1}K_r\tau_m}{1+K_1K_CK_{m1}K_r}\right)}$$

$$\frac{I_a(S)}{E_1(S)} = \frac{K_{1C}(1+S\tau_m)}{1+S\tau_{m2}} \quad ----(2)$$

Where, $K_{1C} = \dfrac{K_1K_CK_{m1}}{1+K_1K_CK_{m1}K_r} \quad ----(3)$

$$\tau_{m2} = \frac{\tau_{m1}+K_1K_CK_{m1}K_r\tau_m}{1+K_1K_CK_{m1}K_r} \quad ----(4)$$

Since, $K_1K_CK_{m1}K_r >>> 1$, so 1 is neglected in (3) & (4)

So (3) becomes, $K_{1C} = \dfrac{\cancel{K_1K_CK_{m1}}}{\cancel{K_1K_CK_{m1}}K_r} = \dfrac{1}{K_r} \quad ----(5)$

& (4) becomes, $\tau_{m2} = \dfrac{\tau_{m1}+K_1K_CK_{m1}K_r\tau_m}{K_1K_CK_{m1}K_r}$

$$\Rightarrow \tau_{m2} = \frac{\tau_{m1}}{K_1K_CK_{m1}K_r} + \frac{\cancel{K_1K_CK_{m1}K_r}\tau_m}{\cancel{K_1K_CK_{m1}K_r}}$$

$$\Rightarrow \tau_{m2} = \frac{\tau_{m1}}{K_1K_CK_{m1}K_r} + \tau_m \quad ----(6)$$

$\tau_m >>> \tau_{m1}$, so τ_{m1} is neglected in (6)

$\tau_{m2} = \tau_m \quad ----(7)$

Now, subs (5) & (7) in (2)

$$\frac{I_a(S)}{E_1(S)} = \frac{\dfrac{1}{K_r}(1+S\tau_m)}{1+S\tau_m} \Rightarrow \frac{(\cancel{1+S\tau_m})}{K_r(\cancel{1+S\tau_m})} \Rightarrow \frac{1}{K_r} \quad ----(8)$$

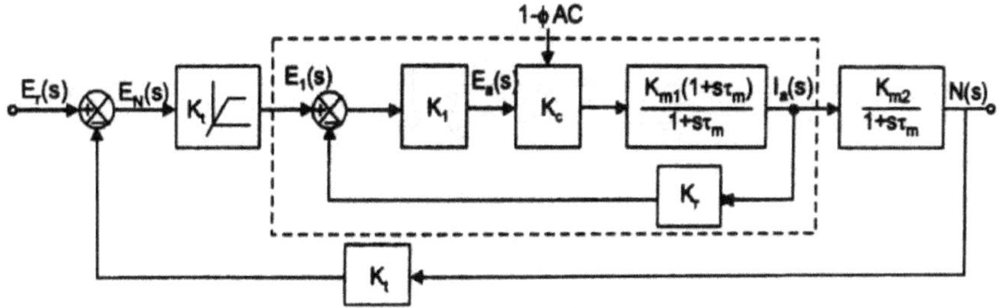

Fig.5.10. Functional Block diagram

Without tacho generator filter

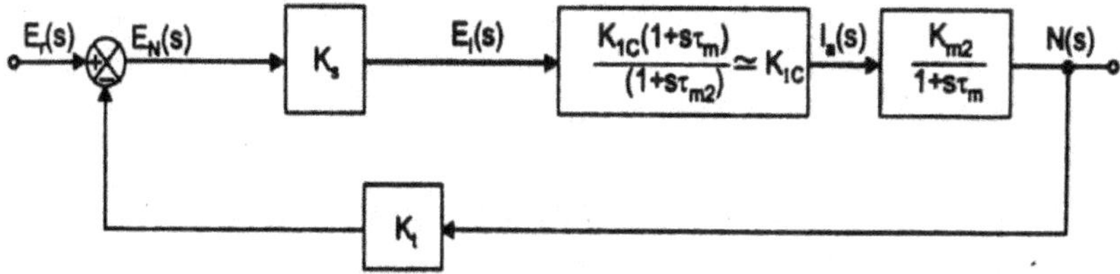

Fig.5.11. Simplified Functional Block diagram

From the simplified block diagram,

$$\frac{N(S)}{E_r(S)} = \frac{\dfrac{K_S K_{1c} K_{m2}}{1+S\tau_{m2}}}{1+\dfrac{K_S K_{1c} K_{m2}}{1+S\tau_{m2}}*K_t}$$

$$\frac{N(S)}{E_r(S)} = \frac{\dfrac{K_S K_{1c} K_{m2}}{1+S\tau_{m2}}}{\dfrac{1+S\tau_{m2}+K_S K_{1c} K_{m2} K_t}{1+S\tau_{m2}}}$$

$$\frac{N(S)}{E_r(S)} = \frac{K_S K_{1c} K_{m2}}{1+S\tau_{m2}+K_S K_{1c} K_{m2} K_t}$$

$$\frac{N(S)}{E_r(S)} = \frac{K_S K_{1c} K_{m2}}{(1+K_S K_{1c} K_{m2} K_t)+S\tau_{m2}} \qquad ----(9)$$

Divide $(1+K_S K_{1c} K_{m2} K_t)$ in both numerator & Denominator in (9)

$$\frac{N(S)}{E_r(S)} = \frac{\dfrac{K_S K_{1c} K_{m2}}{1+K_S K_{1C} K_{m2} K_t}}{\left(\dfrac{1+K_S K_{1C} K_{m2} K_t}{1+K_S K_{1C} K_{m2} K_t}\right) + \dfrac{S\tau_{m2}}{1+K_S K_{1C} K_{m2} K_t}}$$

$$\frac{N(S)}{E_r(S)} = \frac{\dfrac{K_S K_{1c} K_{m2}}{1+K_S K_{1C} K_{m2} K_t}}{1+\dfrac{S\tau_{m2}}{1+K_S K_{1C} K_{m2} K_t}}$$

$$\frac{N(S)}{E_r(S)} = \frac{K_2}{1+S\tau_2} \quad ----(10)$$

Where, $K_2 = \dfrac{K_S K_{1c} K_{m2}}{1+K_S K_{1C} K_{m2} K_t} \quad ----(11)$

$$\tau_2 = \frac{\tau_{m2}}{1+K_S K_{1C} K_{m2} K_t} \quad ----(12)$$

Since, $K_t K_s K_{1C} K_{m2} >>> 1$, therefore 1 is neglected in (11) & (12)

(11) becomes, $K_2 = \dfrac{\cancel{K_S K_{1C} K_{m2}}}{\cancel{K_S K_{1C} K_{m2}} K_t} = \dfrac{1}{K_t} \quad ----(13)$

$$\tau_2 = \frac{\tau_{m2}}{K_S K_{1C} K_{m2} K_t} \quad ----(14)$$

WKT,

$$\frac{I_a(S)}{E_r(S)} = \frac{N(S)}{E_r(S)} * \frac{I_a(S)}{N(S)}$$

$$\frac{I_a(S)}{E_r(S)} = \frac{K_2}{1+S\tau_2} * \frac{1+S\tau_m}{K_{m2}}$$

Where, $\dfrac{I_a(S)}{N(S)} = \dfrac{1+S\tau_m}{K_{m2}} \quad ----(15)$

$$\frac{I_a(S)}{E_r(S)} = \frac{K_2(1+S\tau_m)}{K_{m2}(1+S\tau_2)} \quad ----(16)$$

During load change, speed error is large. Speed equation from (15) will be,

$$(15) => \frac{N(S)}{I_a(S)} = \frac{K_{m2}}{1+S\tau_m}$$

$$N(S) = I_a(S) \cdot \frac{K_{m2}}{1+S\tau_m} \quad ----(17)$$

For step input (17) becomes,

$$N(S) = \frac{I_a}{S} \cdot \frac{K_{m2}}{1+S\tau_m} \quad ----(18)$$

With tacho generator filter

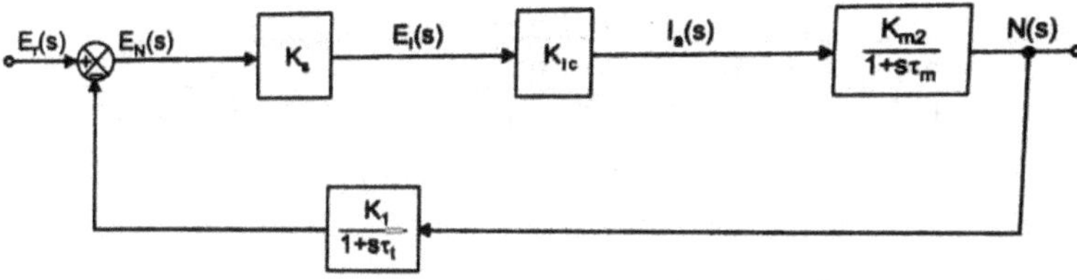

Fig.5.12. Functional Block diagram with added tacho generator filter From the functional block diagram with added tacho generator filter,

$$\frac{N(S)}{E_r(S)} = \frac{\dfrac{K_S K_{1c} K_{m2}}{1+S\tau_m}}{1+\dfrac{K_S K_{1C} K_{m2}}{1+S\tau_m} * \dfrac{K_t}{1+S\tau_t}}$$

$$\frac{N(S)}{E_r(S)} = \frac{\dfrac{K_S K_{1c} K_{m2}}{1+S\tau_m}}{1+\dfrac{K_S K_{1C} K_{m2} K_t}{(1+S\tau_m)(1+S\tau_t)}}$$

$$\frac{N(S)}{E_r(S)} = \frac{\dfrac{K_S K_{1c} K_{m2}}{\cancel{1+S\tau_m}}}{\dfrac{(1+S\tau_m)(1+S\tau_t)+K_S K_{1C} K_{m2} K_t}{\cancel{(1+S\tau_m)}(1+S\tau_t)}}$$

$$\frac{N(S)}{E_r(S)} = \frac{K_S K_{1c} K_{m2}}{\dfrac{(1+S\tau_m)(1+S\tau_t)+K_S K_{1C} K_{m2} K_t}{(1+S\tau_t)}}$$

$$\frac{N(S)}{E_r(S)} = \frac{K_S K_{1c} K_{m2}(1+S\tau_t)}{(1+S\tau_m)(1+S\tau_t)+K_S K_{1C} K_{m2} K_t}$$

$$\frac{N(S)}{E_r(S)} = \frac{K_S K_{1c} K_{m2}(1+S\tau_t)}{(1+S\tau_m + 1+S\tau_t + S^2\tau_m\tau_t) + K_S K_{1C} K_{m2} K_t}$$

$$\frac{N(S)}{E_r(S)} = \frac{K_S K_{1c} K_{m2}(1+S\tau_t)}{(1+K_S K_{1C} K_{m2} K_t) + S(\tau_m+\tau_t) + S^2\tau_m\tau_t} \ ----(19)$$

Divide $(1+K_t K_S K_{m2} K_{1C})$ in numerator and denominator of (19)

$$\frac{N(S)}{E_r(S)} = \frac{\dfrac{K_S K_{1c} K_{m2}(1+S\tau_t)}{1+K_t K_S K_{m2} K_{1C}}}{\dfrac{1+K_S K_{1C} K_{m2} K_t}{1+K_t K_S K_{m2} K_{1C}} + \dfrac{S(\tau_t+\tau_m)}{1+K_t K_S K_{m2} K_{1C}} + \dfrac{S^2\tau_m\tau_t}{1+K_t K_S K_{m2} K_{1C}}}$$

$$\frac{N(S)}{E_r(S)} = \frac{\dfrac{K_S K_{1c} K_{m2}(1+S\tau_t)}{1+K_t K_S K_{m2} K_{1C}}}{1 + \dfrac{S(\tau_t+\tau_m)}{1+K_t K_S K_{m2} K_{1C}} + \dfrac{S^2\tau_m\tau_t}{1+K_t K_S K_{m2} K_{1C}}}$$

$$\frac{N(S)}{E_r(S)} = \frac{K_S K_{1c} K_{m2}(1+S\tau_t)}{\left[1 + \dfrac{S(\tau_t+\tau_m)}{K^l} + \dfrac{S^2\tau_m\tau_t}{K^l}\right](1+K_t K_S K_{m2} K_{1C})} \ ---(20)$$

Where, $K^l = 1+K_t K_S K_{m2} K_{1C}$

$K_t K_S K_{m2} K_{1C} >>> 1$, so 1 is neglected

So, $K^l = K_t K_S K_{m2} K_{1C}$

WKT,

$$\frac{I_a(S)}{E_r(S)} = \frac{N(S)}{E_r(S)} * \frac{I_a(S)}{N(S)} \ ----(21)$$

$$\frac{I_a(S)}{N(S)} = \frac{1+S\tau_m}{K_{m2}} \ ----(22)$$

Subs (20) & (22) in (21)

$$\frac{I_a(S)}{E_r(S)} = \frac{K_S K_{1c} K_{m2}(1+S\tau_t)}{\left[1 + \dfrac{S(\tau_t+\tau_m)}{K^l} + \dfrac{S^2\tau_m\tau_t}{K^l}\right](1+K_t K_S K_{m2} K_{1C})} * \frac{1+S\tau_m}{K_{m2}}$$

$$\frac{I_a(S)}{E_r(S)} = \frac{K_S K_{1C}\cancel{K_{m2}}(1+S\tau_t)}{\left[1+\frac{S(\tau_t+\tau_m)}{K^l}+\frac{S^2\tau_m\tau_t}{K^l}\right](1+K_t K_S K_{m2} K_{1C})} * \frac{1+S\tau_m}{\cancel{K_{m2}}}$$

$$\frac{I_a(S)}{E_r(S)} = \frac{K_S K_{1C}(1+S\tau_t)(1+S\tau_m)}{\left[1+\frac{S(\tau_t+\tau_m)}{K^l}+\frac{S^2\tau_m\tau_t}{K^l}\right](1+K_t K_S K_{m2} K_{1C})} \quad ----(23)$$

b) PI Controller

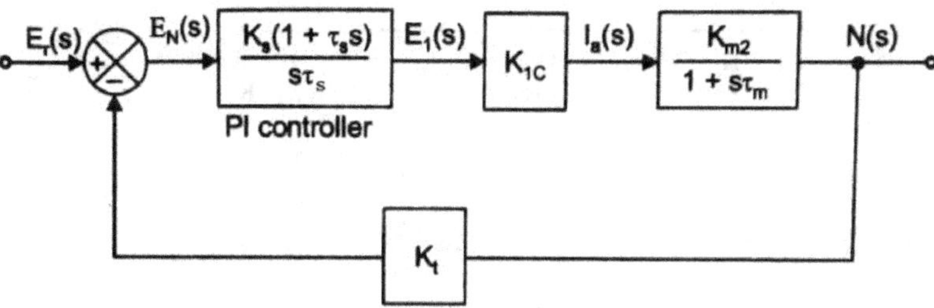

Fig.5.13. Speed Control Loop with PI controller An integral feedback is used to eliminate the steady state error.

Transfer function of PI controller is, $\dfrac{K_S(1+S\tau_S)}{S\tau_S}$

From the Speed Control Loop with PI controller block diagram,

$$\frac{N(S)}{E_r(S)} = \frac{\dfrac{K_S(1+S\tau_S)K_{1C}K_{m2}}{S\tau_S(1+S\tau_m)}}{1+\dfrac{K_S(1+S\tau_S)K_{1C}K_{m2}}{S\tau_S(1+S\tau_m)}*K_t} \quad ----(24)$$

$$\frac{N(S)}{E_r(S)} = \frac{\dfrac{K_S K_{1C}K_{m2}(1+S\tau_S)}{\cancel{S\tau_S(1+S\tau_m)}}}{\dfrac{S\tau_S(1+S\tau_m)+K_S(1+S\tau_S)K_{1C}K_{m2}K_t}{\cancel{S\tau_S(1+S\tau_m)}}}$$

$$\frac{N(S)}{E_r(S)} = \frac{K_S K_{1C}K_{m2}(1+S\tau_S)}{S\tau_S(1+S\tau_m)+K_S K_{1C}K_{m2}K_t(1+S\tau_S)}$$

$$\frac{N(S)}{E_r(S)} = \frac{K_S K_{1C}K_{m2}(1+S\tau_S)}{S\tau_S+S^2\tau_S\tau_m+K_S K_{1C}K_{m2}K_t+S\tau_S K_S K_{1C}K_{m2}K_t}$$

$$\frac{N(S)}{E_r(S)} = \frac{K_S K_{1C} K_{m2}(1+S\tau_S)}{K_S K_{1C} K_{m2} K_t + S\tau_S(1+K_S K_{1C} K_{m2} K_t) + S^2 \tau_S \tau_m} \quad\text{----(25)}$$

Divide $K_S K_{1C} K_{m2} K_t$ in numerator & denominator of (25)

$$\frac{N(S)}{E_r(S)} = \frac{\dfrac{\cancel{K_S K_{1C} K_{m2}}(1+S\tau_S)}{\cancel{K_S K_{1C} K_{m2}} K_t}}{\dfrac{\cancel{K_S K_{1C} K_{m2} K_t}}{\cancel{K_S K_{1C} K_{m2} K_t}} + \dfrac{S\tau_S(1+K_S K_{1C} K_{m2} K_t)}{K_S K_{1C} K_{m2} K_t} + \dfrac{S^2 \tau_S \tau_m}{K_S K_{1C} K_{m2} K_t}}$$

$$\frac{N(S)}{E_r(S)} = \frac{\dfrac{(1+S\tau_S)}{K_t}}{1 + \dfrac{S\tau_S(1+K_S K_{1C} K_{m2} K_t)}{K_S K_{1C} K_{m2} K_t} + \dfrac{S^2 \tau_S \tau_m}{K_S K_{1C} K_{m2} K_t}} \quad\text{----(26)}$$

$K_S K_{1C} K_{m2} K_t >>> 1$, so neglect 1 in $(1 + K_S K_{1C} K_{m2} K_t)$ of (26)

So (26) becomes,

$$\frac{N(S)}{E_r(S)} = \frac{\dfrac{(1+S\tau_S)}{K_t}}{1 + \dfrac{S\tau_S \cancel{K_S K_{1C} K_{m2} K_t}}{\cancel{K_S K_{1C} K_{m2} K_t}} + \dfrac{S^2 \tau_S \tau_m}{K_S K_{1C} K_{m2} K_t}}$$

$$\frac{N(S)}{E_r(S)} = \frac{\dfrac{(1+S\tau_S)}{K_t}}{1 + S\tau_S + S^2 \tau_S \tau_2} => \frac{(1+S\tau_S)}{K_t(1+S\tau_S+S^2 \tau_S \tau_2)} \quad\text{----(27)}$$

Where, $\tau_2 = \dfrac{\tau_m}{K_S K_{1C} K_{m2} K_t}$

WKT,

$$\frac{I_a(S)}{E_r(S)} = \frac{N(S)}{E_r(S)} * \frac{I_a(S)}{N(S)} \quad\text{----(28)}$$

$$\frac{I_a(S)}{N(S)} = \frac{1+S\tau_m}{K_{m2}} \quad\text{----(29)}$$

Substitute (27) & (29) in (28)

$$\frac{I_a(S)}{E_r(S)} = \frac{(1+S\tau_S)}{K_t(1+S\tau_S+S^2 \tau_S \tau_2)} * \frac{1+S\tau_m}{K_{m2}}$$

$$\frac{I_a(S)}{E_r(S)} = \frac{(1+S\tau_S)(1+S\tau_m)}{K_t K_{m2}(1+S\tau_S+S^2 \tau_S \tau_2)} \quad\text{----(30)}$$

5.6. CONVERTER SELECTION AND CHARACTERISTICS

Selection Factors

- Rating of the converter
- Power switches usage
- Motor load specification

Controller Characteristics

Let the maximum current of the motor be I_{max}.

RMS value of current is, $I_{RMS} = \dfrac{I_{max}}{\sqrt{3}} = 0.577\ I_{max}$.

Here, voltage rating would be maximum

$$V_t = \sqrt{2}\ V$$

I_1 is the fundamental RMS component of the input AC current,

$$I_1 = \frac{1}{\sqrt{2}}\frac{2\sqrt{3}}{\pi}\ I_{max} = \frac{\sqrt{3}\sqrt{2}}{\pi}\ I_{max} = 0.78\ I_{max}$$

Output power of the converter,

$$P_o = V_a\ I_{max}$$
$$= (1.35\ V\ Cos\alpha)\ I_{max}$$
$$= 1.35\ V\ I_{max}\ Cos\alpha$$

Neglecting the losses in converter, the input power becomes equal to output power.

So,

$$P_o = P_i = 1.35\ V\ I_{max}\ Cos\ \alpha$$
$$= \sqrt{3}\ V\ I_1\ Cos\ \alpha$$

Reactive power $Q_i = 1.35\ V\ I_{max}\ Sin\ \alpha$

$$= \sqrt{3}\ V\ I_1\ Sin\ \alpha$$

α - Power factor angle

Input apparent power is

$$P_{VA} = \sqrt{(P_i)^2 + (Q_i)^2} = 1.35\ V\ I_{max} = \sqrt{3}\ V\ I_1$$